CW01211999

EUROPE'S EXPERIMENT IN FUSION
The JET Joint Undertaking

$$^2_1D + ^3_1T \rightarrow ^4_2He + ^1_0n + 17.6 \text{ MeV}$$

EUROPE'S EXPERIMENT IN FUSION

The JET Joint Undertaking

E. N. SHAW

1990

NORTH-HOLLAND
AMSTERDAM • OXFORD • NEW YORK • TOKYO

© Elsevier Science Publishers B.V., 1990

All rights reserved. No part of this publication may be reproduced, stored in a retrieval system, or transmitted in any form or by any means, electronic, mechanical, photocopying, recording or otherwise, without the prior written permission of the Publisher, Elsevier Science Publishers B.V., P.O. Box 211, 1000 AE Amsterdam, The Netherlands.

Special regulations for readers in the U.S.A. - This publication has been registered with the Copyright Clearance Center Inc. (CCC), Salem, Massachusetts. Information can be obtained from the CCC about conditions under which photocopies of parts of this publication may be made in the U.S.A. All other copyright questions, including photocopying outside of the U.S.A., should be referred to the Publisher.

No responsibility is assumed by the Publisher for any injury and/or damage to persons or property as a matter of products liability, negligence or otherwise, or from any use or operation of any methods, products, instructions or ideas contained in the material herein.

ISBN: 0 444 88330 4

North-Holland
ELSEVIER SCIENCE PUBLISHERS B.V.
P.O. Box 211
1000 AE Amsterdam
The Netherlands

Sole Distributors for the U.S.A. and Canada:
ELSEVIER SCIENCE PUBLISHING COMPANY, INC.
655 Avenue of the Americas
New York, N.Y. 10010
U.S.A.

```
          Library of Congress Cataloging-in-Publication Data

Shaw, E. N.
    Europe's experiment in fusion : the JET Joint Undertaking / E.N.
  Shaw.
       p.    cm.
    ISBN 0-444-88330-4
    1. Controlled fusion--Research--Europe.  2. Fusion reactors-
  -Research--Europe.  3. Plasma confinement--Research--Europe.  4. JET
  Project.   I. Title.
  QC791.76.E85S53  1990
  621.48'4'07204--dc20                                      89-48998
                                                                 CIP
```

PRINTED IN THE NETHERLANDS

CONTENTS

Foreword	vii
Preface	ix
Fusion	1
First steps towards collaboration in fusion	5
The tokamak emerges	13
The Enriques study	19
Organisational studies	25
Design phase opens	33
JET agreement	59
Enlargement of the fusion community	79
Interim phase	83
Management	107
The JET device	113
Forward planning	131
Funding	137
Project control	141
Development problems	145
Construction and commissioning	153
Transition to operation	169
Operation and inauguration	177
List of principal industrial suppliers	183
Name index	187
Subject index	189

FOREWORD

Europe in the late eighties is turning its attention more and more to the Single European Act due to enter into force in 1992. It is a measure of the achievement of the Communities that we have progressed so far. This unification, limited though it may be, has become possible only because men and women in the different countries of Europe have been seeking patiently over the past thirty years to define those areas of activity where national practices can be harmonised and joint endeavours promoted.

Thermonuclear fusion research into magnetically confining a plasma of very high temperature is one of the outstanding examples of cooperation within the Communities, a cooperation in no way defensive, as Sweden and Switzerland are full participants and exchanges with Japan, the USA and the USSR, and indeed all countries with fusion programmes, are completely open.

In JET we have the very first Joint Undertaking created by the Communities. Its evident technical success is the result of a dynamic involvement of all the major fusion laboratories in western Europe. In these coming years, even if difficulties have to be overcome before the Single European Act is fully implemented, people can take heart from the knowledge that when it is determined, Europe is well able to create enterprises of the highest class. JET is an inspiration as well as a technological triumph.

J. Teillac

PREFACE

Towards the end of the first phase of construction of the Joint European Torus, the Director of the Project, Dr Hans-Otto Wüster, invited me to record the evolution of this unique enterprise of the European Communities. JET was the first Joint Undertaking to be created *ab initio* within the Communities and he was anxious that its progress should not go unchronicled. Whilst not dissimulating his personal pride in the Project, nor the pleasure he anticipated from recalling its history, his primary interest was to see put on record those aspects of the Joint Undertaking which might be relevant to other European ventures and to see proper credit given to the international team which has realised the Project. It was not a panegyric that he looked for, but an independent evaluation wherein the author had full liberty to define his priorities and to offer comment that was strictly his own.

On this understanding I was honoured to accept the invitation, accompanied as it was by the offer of complete cooperation by the Project. This cooperation has been more than merely formal in the sense of access being given to written records. Many people within JET itself, within the Commission and within the Associations in the different countries have provided data and shared their personal reactions to past events. To all these people I should like to express my sincerest thanks and also to those within and outside the Project who read the original draft and pointed out the errors it contained.

Direct quotation of private communications has for the most part been avoided; this book is rather a distillation of the written and verbal information that has been gathered together. Specific references are also difficult to cite, as for the most part the sources are unpublished. The minutes of internal meetings are, for example, not generally accessible, nor evidently the personal files of those involved.

At the outset it was intended to review the development of the Project from the time the decision to proceed was taken, but it became clear that only in the context of the emergence of the tokamak as the leading fusion configuration and the elaboration of the specifications of the JET machine could the Joint Undertaking be viewed in perspective. A later decision was to limit the period covered to that leading up to the official inauguration. JET is a continuing Project and this marked the end of that phase of the Joint Undertaking where the basic objectives had been achieved.

This book is about JET. It does not attempt to review the development of fusion devices in general or even tokamaks world-wide. Only when they had a direct impact on JET are events in other continents evoked. Moreover, only the first few months of operation are considered whereas it is the experimental programme that is of the essence.

Few people outside the immediate fusion environment are aware of the degree of cooperation that has been established in western European fusion research. All the major centres including those in Sweden and Switzerland participate fully in the Communities' joint fusion programme. In this, JET fulfils two main functions. As the largest and most performant of all the tokamaks, it adds a scaling dimension to other experiments and as a joint undertaking, it brings together physically as well as morally all the fusion laboratories of western Europe.

The state of fusion research is also not well understood outside the laboratories directly involved, in part because of historical associations with fission, in part because it is a complex subject, but also in part because the fusion community has tended to create a false image by implying that illimitable sources of clean power were almost within our grasp. Thermonuclear fusion is a field of research where there is certainly an ultimate applied objective — a reacting plasma producing economic power — and where feasibility has already been demonstrated — the Sun shines. But that goal should be seen as a distant summit not an immediate peak, to be scaled by frontal assault. The winding path may prove to be more effective in the long term. With each step that is taken it is clear that we need much more information about the terrain, in other words a much greater knowledge of the elementary physics of plasma behaviour. This is not entirely evident in the arguments advanced at the times when continued funding is being sought. Boldness is, of course, a necessary ingredient of successful research but then so is detailed measurement, theoretical understanding, academic insight and the irreverent injection of ideas from the next generation.

Plasma physics is a major field of science in its own right, as challenging intellectually as any of the more traditional disciplines. Its pursuit requires a range of devices, some of them big. These demand innovative engineering of the highest quality. They are nevertheless still quite modest in comparison with those needed for high energy physics, astronomy and synchrotron radiation research, not to mention space. Even without the prospect of fusion power stations one day, the European fusion programme with a current total expenditure of 450 million ECU *per annum* (including 110 million ECU on JET) could be justified by its inherent scientific interest, particularly if the research could become more broadly based with the Universities more directly involved. One can note that the annual budget of CERN alone is greater, and this forms only a part of the high energy physics effort in Europe.

When the Joint Undertaking was set up, the objective was to build and operate for a limited time a device whose specification was largely defined in 1975. When Dr Wüster accepted the role of Director of the Project he make it a matter of personal honour to see that that objective would be fulfilled. It is a tragedy that this history should honour his memory rather than the first phase in the crowning achievement of his career. It is no exaggeration to say that when he died from a heart attack on 30 June 1985 at the age of 58, he did so in the service of the Joint Undertaking. He drove himself to the limit and finally that limit was exceeded.

But JET goes on, now under the leadership of Paul-Henri Rebut, the inspiration behind the JET device. The machine is today much changed from the 1975 specification upon which the agreement to proceed was based and our knowledge of plasma behaviour under conditions not previously explored is that much further forward. This is not to say that now we understand. Nature continues to surprise us. But then surprise and discovery are the very breath of science, a mainspring of human culture not just technology.

In the Joint Undertaking, Europe has its central pivot for fusion research, a core of experience that is second to none in the world. The enthusiasm and dedication that marked the Project from the beginning of the Design Phase continue undimmed. When the Communities discuss the future programme, care must be taken to see that what has been built up over so many years is not sacrificed to some short-term economy measure or to satisfy precepts that are now out of date. In JET, Europe has created a valuable resource in people as well as machines. The way that resource is husbanded will be a test of the long-term credibility of Community undertakings in general.

E N Shaw

FUSION

If, in the 1970's, Europe was to continue to make a major contribution to understanding how the fusion of hydrogen atoms at high temperature could be tamed, both scientifically and organisationally, JET was the next step to take. Experience over the previous 20 years had led to the conclusion that of all the many thermonuclear devices studied, one particular form — the tokamak — held out the most promise, and it might be possible to achieve the right conditions in a large machine. Such a machine would be at or beyond the limit of funding that fusion could command in the different European countries, where the trend was to restrict science budgets and to concentrate at one end of the scale on fundamental research with no forseeable practical application and, at the other, on projects liable to show a rapid return on investment. Coincidentally, the moves for Britain to join the European Communities were meeting less opposition and the way would then be clear for the country with one of the strongest fusion traditions to merge its efforts with those of its neighbours into a firmly coordinated programme, the basis for which had been laid by the European Communities, notably Euratom, over a dozen years.

Big fusion devices surround us and one is the source of our very existence — the Sun. While the detailed mechanism of all the reactions that take place at different depths remains a subject for debate even today, the essential information that the main source of heat and light in the Universe comes from the fusion of hydrogen to form helium has been appreciated ever since the elementary structure of the atom was unravelled in the early 1930's. Such fusion can be created in the laboratory using quite modest electrical equipment and the basic data concerning the reactions that are involved have been known for many years.

Simple though it may be to induce nuclear fusion by accelerating particles in a high voltage machine and causing them to strike an appropriate target, the energy released is but a small fraction of the energy needed to drive the machine. In the Sun, everything is so hot, high speed collisions occur automatically through the natural agitation of the atomic particles and the fusion process is called in consequence thermonuclear fusion. Gravity keeps the mass together and prevents the particles from blowing apart. Because the agitation is so vigorous, all the atoms are stripped down into their basic components and form a plasma of electrons and ions in perpetual ferment. This is the natural state of matter in the Universe, the condensation into the familiar atoms and molecules of our own world being a comparatively rare occurrence.

This being so, the interest in studying the plasma state of matter as a scientific subject needs no further explanation and over the past 40 years our knowledge has deepened greatly. Advances in technology dramatically widened the opportunities for experiment in the immediate post-war years, and plasma physics became a subject in its own right. At the same time, the explosive possibilities of fusion were developed while in parallel, the first serious thoughts were given to producing and confining in a controlled and stable manner, plasma hot enough to initiate fusion reactions. Because of the possible military significance, these were for a time kept strictly secret, but the curtain was parted in April 1956 when Academician Igor Kurchatov volunteered to lecture at the Harwell research establishment of the UK Atomic Energy Authority (AEA) on Russian work in the field. Even if general declassification did not happen immediately, publications began to appear in the open literature and early the next year in Britain the Physical Society published a group of scientific papers on thermonuclear reactions. These included the paper by J. D. Lawson in which, for the first time, the minimum conditions for power production were presented in terms of the product of plasma density and confinement time — an expression now universally quoted as the Lawson criterion. Implicit also in the criterion is a temperature high enough to encourage fusion when two particles do collide, i.e. in the region of 100 million degrees Celsius for the optimum mixture of hydrogen fuel.

It was plain by then that the science of fusion confinement and hydrogen bombs had little to do with each other and secrecy for military reasons was no longer justified. Other barriers had nevertheless still to be overcome. In the countries most heavily engaged in fusion studies, research was undertaken within the establishments of the atomic energy agencies where the obsession with commercial secrecy was becoming every bit as restricting as the military. Fortunately the demands of national prestige worked in opposition. Early in 1958 the AEA gave a press conference at Harwell, in parallel with the publication of a scientific paper on the production and retention of a high temperature plasma in a circular machine called ZETA, Zero Energy Thermonuclear Assembly, during which the Harwell Director, Sir John Cockcroft, allowed himself to be persuaded into saying that he was 90% certain that the reaction products observed had a thermonuclear origin (a claim that soon turned out to be false). Cockcroft was far too skilled in manipulating the press for that admission to have been made other than deliberately. It fuelled the media's contention that the US had been muzzling British scientists because they were so much 'in the lead' (gained, it seemed despite the collaboration agreement that the UK and USA had been operating since the end of 1956). This may

Sir John Cockcroft talking to the press during the much reported visit to AERE Harwell early in 1958 when British work on fusion was unveiled. Inset is R. Sebastian Pease on that same occasion.

have had some influence on the Americans who were agonising over what should be the main thrust of their exhibition at the 2nd Geneva Conference on the Peaceful Uses of Atomic Energy due to be held in September of that year. Their main concern was to reinforce their come-back in international scientific prestige following the launch of their first satellite on 31 January 1958—mortifyingly four months after the first Russian sputnik.

Their choice fell on fusion and visitors to the exhibition in Geneva that accompanied the conference, were treated to a dazzling display illustrating the wide range of devices that the Americans had been studying as possible ways of bottling up high temperature plasma. Within the conference hall, fusion held the limelight as declassification permitted the extent of the scientific work in the main countries to be revealed. From that time on, information flowed progressively more freely. Moreover, scientific realism took the place of wishful thinking. Whereas, at the first Geneva conference held three years previously, the President, Homi Bhaba, had been citing a period of 20 years as being all that was necessary to bring fusion into commercial exploitation, by 1958 the magnitude of the basic problem of plasma control was already being appreciated and the end of the century was more the time-scale that people had in mind.

FIRST STEPS TOWARDS COLLABORATION IN FUSION

Inherent to the problem of plasma studies is the difficulty of first generating a plasma at any sizeable temperature and then retaining it for a period long enough for significant measurements to be made. Theory has to rely heavily on experiment, yet experiments can only be designed on theoretical assumptions. This is true to some extent with any science, but whereas in fission, for example, the measurement of certain basic properties of uranium and graphite enabled Fermi in 1942 to build with reasonable confidence a (zero power) reactor that would then operate in a steady state, even producing plasmas that were more than a mere flash was a largely empiric process. In the 1950's, theory was in an elementary state, the length of time high temperature plasma discharges could be held was measured in millionths of a second and diagnostic instruments were crude. The laser that is now so important as a measuring tool, for example, did not make its appearance until 1960. Our knowledge has grown enormously over the past 20 years but the basic difficulty of production, stability and measurement, and then interpretation remains, even though the basic interactions are understood and the relevant numbers are well established. An analogy can be made with metallurgy where, despite the wealth of understanding that has been amassed on the way atoms stick together, it is still extremely difficult to custom-design a new material. Metallurgy, however, has a few centuries of observation behind it and deals with materials that are (generally) stable, manageable and available in convenient quantities. Plasma research is concerned with stuff that is highly unstable, extremely elusive and available for only very short periods of time in largely inaccessible containers.

Against this daunting image, 1958 marked the time when plasma research shook free of its security bonds and became a field of study both academic and applied that no country with pretensions in the science/technology stakes could ignore, for behind the academic interest and the traditional wish for scientists to shine before their peers, was the nagging fear that a dramatic break-through could overnight give commercial dominance to those who understood the science. The new status of thermonuclear research was reflected in the structural changes that took place in Europe.

In Germany, for example, Heisenberg's moves to establish a national effort were translated into the creation of the Max-Planck-Institut für Plasma Physik (IPP) at Garching, near Munich. Academic activity had been signalled by the publication in 1957 by Arnulf Schlüter et al. of work at the Astrophysics Division of the MPI in Göttingen. Heisenberg initially considered the possibility of establishing a group in Karlsruhe and then at the Technical University in Munich, but was influenced by what he learned as Vice-President of the Council of CERN, the European Organisation for Nuclear Research in Geneva that had been set up in 1954 to do fundamental research on elementary particles using high energy accelerators. Aware of the coming explosion in fusion information, the CERN Council had approved in June 1958 the setting up of a Study Group on plasma physics to evaluate the research programmes then in progress or preparation, and to consider ways of facilitating fusion research work in Europe. This Group under the chairmanship of John Adams (at that time Director of the Division responsible for building CERN's 'big machine') reported in May of the following year. Its recommendations, accepted by the Council, were that it would not be appropriate to set up a new European research laboratory at that stage and CERN should continue to sponsor the Study Group while individual countries evaluated their own needs. Heisenberg became convinced of the need for Germany to have its own special centre and IPP was established, provisionally as a limited liability company with two shareholders, the Max-Planck-Gessellschaft and Heisenberg himself. This holding was modified in December 1970, but it anticipated the unusual position that IPP would have in the general pattern of Max-Planck Institutes. Apart from being the largest, it is project orientated and now is 90% financed by the Federal Government and only 10% by the Bavarian Government.

Soon afterwards a second fusion centre was established at the Kernforschungsanlage (KFA) Jülich near Düsseldorf. KFA is a largely national centre with a number of independent institutes grouped around a high temperature gas cooled reactor project (the 'pebble bed'). Although the fusion effort is much smaller than that of IPP which through the 60's had a very broad programme of research, the KFA team was well above the threshold level and had its own energetic programme.

In the Netherlands, thermonuclear work that had begun in 1957 under the Stichting voor Fundamenteel Onderzoek der Materie (FOM) was, in November of the following year, moved to the newly acquired castle of Rijnhuisen at Jutphaas (now renamed Nieuwegein), one of the most attractive sites for a research establishment to be found in the plasma field. A charming moated manor house provides the central offices and working area and additional laboratories have been added over the years in the surrounding park. The driving force was the science and the policy of the funding bodies to support quality work centred on gifted individuals almost irrespective of the field. Only later did the possibility of fusion providing an alternative energy source assume any significance.

In Britain, instead of transferring fusion research to the reactor site at Winfrith as had been announced in 1958, the AEA was authorised to set up a new establishment at Culham, a few kilometres away from Harwell, where much of the original work had been carried out until then. Not all the fusion work left Harwell, but from January 1960 when the site was opened, Culham became the principal fusion centre (with John Adams as Director). It was run essentially as an open laboratory and soon was regarded by the AEA and by the staff alike as something apart from the mainstream of atomic energy activities, even though still subject to the AEA's remit of doing orientated research only; fundamental research is not allowed unless it clearly relates to the exploitation of nuclear energy.

In France, the change was less evident. Fusion research had been established at one of the early sites of the Commisariat à L'Energie Atomique (CEA) at Fontenay-aux- Roses and there it remained as the newer establishments at Saclay, Marcoule and, from early 1960, Cadarache were set up, although some fusion groups were formed in Saclay. Within the very broad programme of research and development that aimed to make France self-sufficient in nuclear power (and formally from 1958 in military) technology, fusion was of minor importance only. Fission had by far the greatest priority and fusion has always been seen primarily as an interesting scientific field, even if at some time it might have practical applications, and is funded through the Département de Recherche Fondamentale. Nevertheless, France reacted to the general mood and its fusion programme expanded.

The year 1958 was also marked by the coming into existence of the European Economic Community and the European Community for Atomic Energy (Euratom), in a political climate somewhat changed from that obtaining when the grand design had been first conceived at Messina in June 1955. Already by March 1957 when the Treaty of Rome was signed, the supranational powers of Euratom had been circumscribed and the following months saw the six Member States being more concerned with entrenchment than with looking for new areas in which to open cooperation. Primarily concerned with the development of power from fission in all its aspects, Euratom instead of becoming the lynch pin around which the atomic energy programmes of the Six revolved, found itself rebuffed particularly by the strongest partner France. The coming to power of General de Gaulle could only deepen the divide which developed. Euratom's initiation was not helped by having a first President (Louis Armand) who was too ill to assume his functions, and the burden of trying to establish a European programme fell on Jules Guéron, the Director-General of Research and Development, whose ambitions on behalf of the Commission found little response in his home country.

Listed in Annex 1 to the Rome agreement, among the areas of research the Commission was required to promote and facilitate within the Member States, was 'the study of fusion relating in particular to the behaviour of an ionised plasma under the influence of electro-magnetic forces and to the thermodynamics of extremely high temperatures'. The man appointed to look after this sector was Donato

Palumbo, previously Professor of Theoretical Physics at the University of Palermo — a quiet, self-effacing man with a keen appreciation of the science and a sensitive awareness of the preoccupations of scientists in the Member States. His long-term vision, tempered by his acute awareness of what was practicable, his persistence in seeking solutions to problems between laboratories and avoiding confrontations have, over the years, made European collaboration in fusion research one of the Communities' greatest achievements, and certainly Euratom's most notable success.

Amongst the leaders of fusion research in Europe were A. Schlüter (left) and C. Braams (right) seen here with D. Palumbo discussing the foundation of JET.

Palumbo saw Euratom's role as supporting research in the national centres that were then developing, rather than mounting a common programme within the Joint Research Centre (JRC) which the Treaty required to be set up. The JRC was, from the beginning, a source of contention: instead of one single centre being created, finally four were recognised. Of these, the least controversial has been the Nuclear Measurements Centre at Geel in Belgium and the most controversial the Ispra centre in northern Italy which was handed over to Euratom in 1959 and became (unlike the two others at Karlsruhe and Petten) totally funded by the Commission. Ispra's history is one of abandoned projects, perpetual squabbles over its forward programme, and labour unrest, all of which has added up to an establishment of low output and low morale. Italy has always felt especially involved and its approach to practically all the Communities' scientific projects has been conditioned at one time or another by its anxiety to obtain continuing support for Ispra, in the face of opposition from its partners.

Palumbo's policy gained little support from Guéron, who preferred a more concrete direction from the Commission, but it found favour with Enrico Medi, the Commissioner in charge of research, a geophysicist given to philosophical abstractions and not a fission man at all. As a result, Palumbo was able to conclude in July 1959 a Contract of Association with the French CEA, whereby the Commission paid out of Euratom funds ⅔ of the cost of the fusion programme covered by the contract, while the CEA paid ⅓. Money was no problem in Euratom at this time, indeed the Commission was unable to spend what had been agreed for the first 5-year programme, wherein some 11 MUC (million units of account) had been set aside for fusion of which 7.5 MUC was allocated to expenditures in the Member States. Far from there being a queue of projects seeking support, there was a veritable dearth.

The following year the Italian Comitato Nazionale per l'Energia Nucleare (CNEN) became a sub-associate through the CEA—Guéron wanted one contract to cover all work of this type—and a small international team of Euratom people was assembled at the Frascati centre outside Rome where Bruno Brunelli had formed a nucleus of researchers to work on plasma physics. Although primarily concerned with the development of fission technology, CNEN acted as the umbrella organisation for some fundamental research such as high energy physics. Fusion was considered to come into a similar category—high technology, some radioactivity and a broad programme of rather basic research was instituted. Some time later fusion work was also started at Padua and at Milan under the Consiglio Nazionale delle Ricerche (CNR) in collaboration with the Universities.

The next laboratory to be approached by Palumbo was IPP, but they were not prepared to consider acting in a subsidiary capacity to the CEA and in 1961 concluded a separate contract with Euratom. So the precedent was established of having individual Contracts of Association with the different fusion laboratories in the Member States. Jülich came next in 1962, followed almost at once by FOM in the Netherlands. In 1969, a Contract was signed with the State of Belgium on behalf of the Ecole Royale Militaire, where a small group under Paul Vandenplas was making theoretical studies on wave-plasma interactions; and finally Contracts were concluded in 1973 with the UK Atomic Energy Authority and the Danish Atomic Energy Commission.

Initially the terms of these contracts were negotiated quite separately so that, for example, the first contract with IPP provided for a ⅓ support from Euratom, as against ⅔ for the CEA. The difference was the result of different attitudes within the laboratories, IPP being wary at first of accepting a greater contribution because it had no wish to compromise its freedom of action. Programmes were established independently in the different laboratories and the notion of a common European programme was rather hypothetical. The contract with FOM, for example, was designed solely to bring money back into the Netherlands; it had no effect on the programme at Jutphaas or even on its budgets as the money simply went into the FOM general fund. Nevertheless communication was established and the requirement to make information available to all the Members effectively countered the tendency to keep the research confidential.

Motivations as well as objectives were diverse in the different European laboratories. In contrast to Britain for example, where Culham was set up to develop fusion for power generation and this was, in Adams' eyes, its essential justification, the accent in Frascati and Jutphaas was more academic, more directed towards plasma physics. In France it was in the nature of a scientific insurance and at IPP an expression of national identity. Through the 60's a wide variety of systems aimed at plasma confinement were under study reflecting both the open nature of the problems to be confronted and the individualism traditionally associated with academic research. As such there was no great need for European coordination. Physicists had their own channels of communication, there were still close connections with the high energy community, and people met through the CERN Study Group and at international conferences; plasma physics was in a healthy state and the equipment needed for good research was not exhorbitant.

With four countries involved in Contracts of Association some measure of uniformity was desirable if only to minimize bickering over who was getting the best return and an informal board of Associates was formed which then became a formally constituted Groupe de Liaison. This was essentially a forum to report on progress and to discuss forward plans. Delegations were not restricted to one member and at meetings there were often arguments between people from the same country—or even the same laboratory. It was, in effect, a scientific club centred on Euratom and with a responsibility for debating the next multi-annual programmes as they came along. However, it had little executive power and soon after its formation a second committee was set up that met more frequently—the Committee of Directors, comprising the Director of Fusion within Euratom and the Directors of the Associated laboratories. These were the people who controlled staff and equipment, who could implement at the working level any collective decisions that were taken and could modify programmes in the light of discussions.

This was the structure through the 60's which survived—or perhaps fed on the growing despondency in fusion circles. The academic scientist could accept with a certain equanimity, even joy, the results coming from the wide variety of machines designed to confine plasmas in magnetic bottles that had been built in Europe, the USA and USSR. Where programmes however were justified on the grounds of future applications, the outlook was less sanguine. Whatever system was tried, Nature responded with a succession of apparently fundamental effects starting with 'Bohm diffusion' which seemed to create an insuperable barrier to achieving plasmas that were hot, dense and stable all at the same time. Whether the machines were toroidal or linear (with magnetic mirrors at each end), as the density or temperature was increased — with the Lawson criterion still several orders of magnitude away — the plasma became unstable and refused to remain confined. It was a period when magnetic configurations became ever more fanciful and fusion scientists accepted that the road to power would be long, arduous and there would be no quick winners.

The fusion community was not very large in the early 60's—less than 650 professionals in Western Europe, about the same in Eastern Europe, about 60% of this number in the USA and a modest group in Japan, under 2000 altogether. The majority worked in a few national research centres, only a minority in University departments. Two distinct sections could be distinguished—the plasma physicists concerned with fundamental research and the fusion engineers wrestling with the problem of plasma generation, confinement and measurement in machines of ever increasing complexity. Nevertheless the plasma/fusion community was a very close community with the members in easy communication, sustaining each other in the common effort to understand the essential elements of plasma behaviour. Divisions there were in terms of detailed approach, but there was a large body of basic science that welded them together.

Within the six Euratom countries, Contracts of Association provided a defence against budget erosion. Directors could point to the income from Euratom as a justification for expenditure at home and although the argument of something for something is less persuasive than something for nothing, it was helpful in stabilizing the level of effort especially in France and Italy. France had a particularly difficult passage in the late 60's. Following the unrest of May 1968, economies were called for in many sectors and the CEA was required to tighten its belt. Moreover, the recommendations that followed the national review of the (fission) atomic energy programme for 1971-75 came down firmly on the side of water reactors that were being promoted by Electricité de France against the uranium graphite series that was sponsored by the CEA and so strongly championed by the head of reactors, Jules Horowitz. The CEA felt betrayed and was required to reorganise much of its research and development. The fusion effort at Saclay was moved to Grenoble and it was intended that the work at Fontenay should follow. For the staff there this meant a complete upheaval of jobs and home, and morale suffered in consequence. Germany, on the other hand, was enjoying unparalleled economic prosperity, and with Brandt's accession to the Chancellorship in 1968, had made clear that it was no longer prepared to be regarded as a second class nation. As if to emphasize this, the Lände and Federal Governments were making generous appropriations to science. The result was that Germany by the end of the 1960's had by far the biggest fusion programme in Western Europe, double that of the UK whereas a few years earlier the position had been reversed.

Culham had no Euratom subsidy and was suffering in the wake of a panel set up in 1966 by the AEA to review future expenditure on fusion, which in its report the following year made the savage recommendation that a cut of 50% over five years be imposed. Culham was at that time costing around £4M p.a. compared with the AEA's total budget of £70M p.a. For the new Head of Laboratory, R. Sebastian Pease, appointed on Adams' elevation to be AEA Member for Research, this was a daunting prospect and although he was able to make the first 10% cut by the transfer of a project connected more with space research, he was forced to reduce drastically the breadth of Culham's programme and to face a future in the position of poor relation, chronically undernourished, searching for a little extra cash by doing research under contract.

The utility of the Contracts of Association and their political acceptability can be gauged from their survival through the various crises in the Communities and, in particular Euratom, during the second half of the 1960's. In April 1965, the separate Councils supervising the work of the Coal and Steel Community, the Economic Community and Euratom were merged into a common Council so that at Ministerial level, Euratom affairs became enmeshed with those of the Common Market. Then in July 1967 the different Commissions were merged, which resulted in Euratom being broken up and redistributed to the point that it was no longer a single entity. In 1965 France had refused to take part in the work of the Council or its various committees which blocked future budget decisions and it was not until mid-1968 that something like order returned. Throughout it all, fusion retained its cohesion and in the reorganisation, became part of the Industry, Technology and Research Directorate with Palumbo still in charge and still patiently working towards his ideal of a common European fusion programme.

THE TOKAMAK EMERGES

Nurturing European cooperation is an uphill task. In fusion as in so many other branches of science, despite the existence of the Groupe de Liaison, the various working groups it formed to look at different approaches to fusion, and the Committee of Directors, European laboratories looked either west or east for their inspiration rather than among themselves. IPP with a specialisation in stellarators (a particular form of toroidal machine) had its closest links with Princeton; Fontenay-aux-Roses, particularly concerned with mirror devices and a solid core torus was most concerned with work at Livermore, whereas Culham which still pressed on with ZETA (another toroidal machine) and its successors, was in touch with the Americans and also the Russians, the latter facilitated by a bilateral agreement concluded in 1959.

This last collaboration was centred around a common interest in toroidal machines in which a heavy current discharge flowed parallel to a toroidal magnetic field. ZETA was of this type. Hydrogen gas contained in a toroidal chamber acted as the single turn secondary of a transformer. On discharging a high current through the primary, current flowed in the secondary. A plasma was created and heated and the current flowing produced a magnetic field that squeezed the plasma in on itself causing further heating. That magnetic field (poloidal because it goes round the minor circumference) was combined with a toroidal field of about the same strength to provide a helical field that should confine the plasma. Although discharges were mostly of short duration, ZETA had rather irregularly shown signs of producing a hot stable plasma for a short time when the discharge was well established. The Russian variant called a tokamak — a word derived from the Russian words for *current + chamber + magnetic field* — had a similar configuration but a toroidal field that was much stronger than the poloidal field, producing thus a helical field with a much slower twist. This was a design that was to transform the whole fusion scene. When the Russians reported to the plasma physics conference at Novosibirsk in 1968 on the results they were obtaining from their tokamak T3 — a device with a toroidal chamber 2m major and 0.4m minor diameter in which great care had been taken over cleanliness they were frankly disbelieved. Nevertheless, they continued to report stable discharges lasting several hundredths of a second and plasma temperatures inferred to be of several million degrees.

Measurement of temperature was still at that time very difficult, but at Culham a new instrument had been developed which detected the very faint light scattered by the plasma when irradiated by a laser beam and from that could be deduced rather directly the temperature of the electrons in the plasma. Culham was invited to try this device at the Kurchatov Institute on the T3 and in 1969 a team plus five tons of equipment flew to Moscow to make the independent assessment. The fusion world was on tenterhooks because if the Russian results were confirmed, the breakthrough had been made and there was no obvious reason why tokamaks could not be scaled up into bigger machines which should have even better performance. The Americans were highly excited, at one period telephoning Culham every day to find out the latest news as they prepared to reorient their own programme.

The expedition was a great success and when the joint paper announcing the results was published towards the end of 1969 the Russian claims were more than vindicated—temperatures of up to 10 million degrees having been recorded over times of 1/10th of a second. Overnight almost, the fusion world switched its attention to tokamaks and plans for similar devices were prepared in fusion laboratories all round the globe.

Tokamak Projects

While these crucial measurements were being made on T3, an International Conference on Nuclear Fusion Reactors was being held at Culham followed by a symposium at which, for the first time, views were being exchanged on the technological problems that would have to be surmounted before a power generating fusion reactor could be built. It was recognised that none of the systems then under intensive study had demonstrated bare feasibility but it was nevertheless an occasion when plasma physicists and reactor engineers could learn more of each other's problems and the fusion community could gossip together. Pease took the opportunity of sounding out Palumbo on the possible interest there would be in closer cooperation between the Communities and Britain. Pease considered that the UK programme made no sense without a larger machine and the AEA was evidently not ready to fund such a project on its own. Palumbo welcomed the approach for the increased experience this would bring as well as first hand knowledge of a system that had been neglected elsewhere in Europe. Fusion research in Britain was respected on the Continent and in general the feeling was that it was preferable to have the British joining in exchanges rather than 'competing' on the doorstep.

In the following year, 1970, discussions were opened on the way in which cooperation could be progressively developed, and in June, Pease gave a presentation in Brussels before Euratom representatives and the Chairman of the Groupe de Liaison, Cornelis Braams, on the British collaboration with the Russians, concluding with the recommendation that Western Europe should pursue the tokamak line by building a large machine as a joint effort. This was a fairly startling proposal. For laboratories on the Continent, tokamaks were a new line of thinking and it seemed highly premature to rush into building something big before confirming and deepening the Russian work. Braams himself was much more orientated towards small scale plasma physics and within FOM, if not in Jutphaas, machine building on any scale was not even contemplated. Fusion was deemed in the Netherlands to have a low scientific output measured on the very coarse scale of papers published per guilder spent and not yet ripe for consideration as an applied science. As a result its budgets had stabilized with a tendency to decline.

Nevertheless the seed was planted and Culham in parallel sounded out the Americans to see whether the UK might collaborate on the tokamak that was being planned at Princeton — by providing the additional heating for example. Convinced now that toroidal devices were the right route, all work on mirrors was dropped. CLEO, a device planned to be operated as either a stellarator or a tokamak was scheduled to be built initially in its tokamak configuration and systems for injecting neutral beams to heat the plasma were developed for trials on it. At the same time serious studies were started on the design of a large tokamak that could be the basis of a Culham or European venture.

France was amongst the first to be seduced by the Russian tokamak results seeing in them the break-through that would allow them to start afresh with a system that held promise of real progress. Despite budget restrictions and the general despondency at the expected move to Grenoble, the fusion community had not retired into its shell and was almost ready to embark on the building at Grenoble of a big 'hard-core' machine 'Superstator' comprising a torus with a central conductor, the current in which could be used to profile the confining fields. It is difficult now to believe that this type of machine could ever be clean enough to work, and even at the time, there were many doubters in the CEA, but such was the brilliance and forcefulness of its designer Paul-Henri Rebut, it very nearly got accepted. The deciding factor may well have been the great skepticism of the Groupe de Liaison and the general resistance expressed to the device being included in the next European programme. As such it was probably the first time that a nationally conceived fusion project had been modified in any major way as a result of discussions at the European level. It was turned down just as President Pompidou, who succeeded de Gaulle in June 1969, was reversing de Gaulle's regionalisation policy and the move to Grenoble from Fontenay was cancelled. Rebut then turned his fire on the tokamak and within a few months had designed and secured agreement to build the Tokamak de Fontenay-Aux-Roses (TFR). This machine was bigger than T3 and was designed to extrapolate the Russian results, afterwards going beyond to make experiments on supplementary heating of the plasma by external means.

In Italy the tokamak fever had also taken hold. Having a very broad programme of research which included laser work and the generation of high magnetic fields

16 The Tokamak Emerges

The TFR machine built in Fontenay-aux-Roses with inset, Dr P-H. Rebut.

by chemical implosion, the physicists returned from Novosibirsk convinced that a greater concentration was needed and tokamaks should be the line. At the next conference in Rome, the decision was taken to change the stellarator that was on the drawing boards into a small tokamak that was rapidly built and which immediately gave better confinement time than they had ever had before. Plans were also laid for a very ambitious machine which would explore the region of high plasma densities and high magnetic fields and which became known as the Frascati Tokamak (FT).

IPP was more reluctant to jump on the bandwagon as its main investment was in stellarators and despite the disappointing results from the C Stellarator at Princeton, remained to be convinced that tokamaks were the right way to go. Nevertheless IPP was not prepared to be left out of the new line and early in 1970 launched its own tokamak project — Pulsator — a refined version of T3. In a very short time, Europe's main effort became concentrated on the tokamak system. Significantly also the Tokamak Advisory Group of the Groupe de Liaison contained from March 1971 a representative from Culham. This was Alan Gibson, recently back from the USA where he had sat on the committee that had recommended the building of the Princeton Large Tokamak (PLT) — a machine with a torus major radius of 1.3m and a projected plasma current of 1.4MA. His mandate from Pease was to push 'La proposition Pease' as it became known and work towards agreement to build a full-scale machine, preferably as a UK project supported by Europe, but alternatively as a European machine with UK participation.

Apart from the periods when the Six had failed to reach agreement on forward programmes, the Community operated in five-year steps and in the programme due to run from 1971-1975, two very important innovations were introduced by Palumbo and accepted by the Commission and Council. The first of these was the designation of priority projects which would qualify for preferential support and would be funded by the Communities to the extent of some 44% as against the 24% that had become the standard practice for work included in the programmes of the Associations. These priority projects would need to be approved by the Groupe de Liaison and in the competition for recognition, the Associates were forced to discuss them in terms of European usefulness. Until then, Contracts of Association had been confidential and the European programme had consisted essentially of national projects drawing funds from a common source, as much to prevent others getting too much as to give needed additional support. Priority ratings by giving prize money to projects of general interest laid the basis for a programme that was European in more than just name.

Equally important was the second innovation — that of mobility contracts whereby Euratom paid for the extra costs incurred by scientists from one Associate going to work in the laboratory of another. A great deal is talked in Europe of the need to increase internal mobility and there are innumerable bi-lateral cultural agreements under which scientific exchanges are nominally encouraged. In practice they are largely ineffective and even at University level interchange between laboratories is infrequent and confined to a few established scientists whose international reputation leads to their being invited to spend time as guest professors. The Fusion Division's initiative has been largely instrumental in promoting a steadily increasing flow of scientists between laboratories at a variety of levels. Since its introduction, the annual payments have increased from about 50,000 ECU to the mid-80's level of 700,000 ECU, paid to over 100 scientists for visits of two weeks to six months, and the administration has been progressively streamlined.

Peculiar to the scheme (of which JET is not a part, having its own separate provisions) is the manner of payment, namely to the parent laboratory of the scientist. This means that Euratom does not have to become involved in the detail; if the host and source laboratories agree to terms and the transfer is approved by the relevant Association Steering Committee, that is all that is necessary. No further authorisations are required and a simple signature in Brussels is sufficient for the money to be paid out. In the beginning, each case was referred to the Groupe de Liaison, but that was eventually seen as unnecessary.

Simplifying the protocol for Britain's participation in the fusion programme of the Six was the prospect of an enlargement of the Communities. Following General de Gaulle's resignation from the Presidency of France, the veto on enlargement had been lifted and negotiations began on the terms of entry for Britain, Denmark, Ireland and Norway. By May 1971, agreement had been reached on the major issues and in November of that year Denis Willson, the Secretary of Culham, opened discussions on the Laboratory establishing formal links, possibly through a special Contract of Association. Also the Director of Culham was invited to attend the meetings of the Committee of Directors.

Come the end of 1971 and the heads of laboratories were contemplating four tokamaks being built or planned in Europe. The most advanced from the point of view of programme was the French TFR, a tokamak with a major torus radius of 98 cm and a projected current in the plasma of 400,000 A (0.4 MA), Garching was to build Pulsator, a 70 cm, 0.1 MA machine, and Frascati the FT, an 82 cm machine with a planned current of 1.2 MA. Most ambitious was the Culham design, centred on a machine of major radius 150 cm that would exploit the phenomenon of superconductivity, *i.e.* the magnetic fields would be generated by coils wound with a material that at very low temperature had zero electric resistance; plasma currents of up to 2 MA were envisaged. Whereas the first three were essentially national physics experiments to explore the behaviour of current carrying plasma under different conditions, the British proposal was for a machine that would establish the conditions needed in a thermonuclear reactor by producing a reacting plasma at least, and in the most optimistic case reaching ignition—the term used as with an ordinary coal fire to describe the point when the energy liberated takes over from the energy used to light it. We might note in passing that the estimated cost of the machine was put at 10 MUC and the building time five years.

THE ENRIQUES STUDY

By the beginning of 1972, even though the majority feeling was that one needed to know much more about tokamaks before being certain that a really big machine was justified, there was a consensus that a joint study should be made of what might constitute a European programme for the longer term. The Tokamak Advisory Group (TAG) had already come out with a clear recommendation that a team of physicists should be formed to plan a big European machine but the Groupe de Liaison was more circumspect. At its meeting in Varenna in October 1971 it decided to create a working group 'to prepare preliminary designs for the various possible concepts of a future European tokamak. These concepts have to be compared, and their technological and financial consequences and the time-scales involved studied.'

This was endorsed by the Committee of Directors the following month who agreed that the Group should consist of two members per Associated Laboratory plus Culham. Henri Luc, a Belgian physicist employed by Euratom and stationed at Fontenay-aux-Roses, was nominated as Chairman, and he would report to both the Groupe de Liaison and the TAG. It was also agreed that steps should be taken to involve Culham in the work of the Groupe de Liaison (GdL) as an Associated Laboratory on the same standing as the others and that its triennial programme for 1973-75 should be discussed with a view to seeing how it could be fitted into the Communities' five year programme for 1971-75. Culham would have liked its design to have served as the basis for the study but it was pointed out that it was just one proposal among many and the object of the study was not to launch a big project but to examine the problems inherent in different approaches while the work went on to see if a big project was indicated at all.

It should be noted here that the possible variants within the tokamak theme, in addition to the torus size and plasma current that have been mentioned so far, are many. The shape of the torus, its minor diameter, the disposition and current carrying power of the coils that produce the plasma and magnetic fields, the time over which a plasma can be maintained, the provisions for additional heating over and above that caused by the plasma current and the type of energy injection to be used, the power sources that will drive all these... And one more variant of great significance, the question of the composition of the plasma. Ordinary hydrogen and the heavier stable isotope deuterium are simple to handle and experiments at sub-fusion temperatures create effectively no radioactivity. Once fusion is the aim, however, then provision has to be made for coping with the neutrons produced and which activate the surrounding apparatus. Moreover, the conditions needed for fusing deuterium nuclei together are much more severe than those needed for fusing deuterium and the radioactive isotope of hydrogen—tritium, so that if a reacting plasma is really the goal, then in addition to the safety problems associated with a system that becomes radioactive and can only be maintained remotely, one has the immediate problems of handling a fuel that is a radioactive gas that is readily absorbed by any surface it touches.

The Joint European Torus Working Group as it later came to be known (JETWG) met for the first time in February 1972 and began their task of preparing a study of the physical and technical aspects of a high current tokamak. Their mandate was broad, but in practice the only way of arriving at a result that was more than a collection of ideas was to polarise on to one objective. There was no time (even if there had been the will) to examine dispassionately all the possibilities. Moreover at the level of Gibson and Rebut, who were members of the Group, the physicists were already clear that what they wanted was a machine, capable of supporting a plasma in which there was significant fusion heating, at least as performant as anything elsewhere in the world. After all the years in the wilderness, the time had arrived for fusion people to make a new bid to find the promised land. This contrast between the attitudes on the one hand of the policy makers at the level of the Groupe de Liaison or the Committee of Directors and later the Supervisory Board, wishing to have alternatives to consider, and on the other, of the working/design groups concentrating on the development of a specific theme was a feature of all the pre-construction phase. For the majority of physicists, although one could argue over the relative importance of different parameters, the 'primary figure of merit' in a tokamak was the plasma current and very quickly the Working Group came to concentrate on a device capable of producing and sustaining a current of 3MA.

Not all the members of the Committee of Directors were so sure that reactor conditions should be the aim in the first instance. Pease was by far the most determined that the machine should demonstrate and study thermonuclear fusion as such, the others in differing degrees were more concerned with understanding plasma behaviour with different types of heating and going over to tritium only when this was necessary. Not all were yet prepared to accept that a big tokamak should be built even. Schlüter, the Chief Scientific Director of IPP, for example, had still a preference for the stellarator configuration and was afraid that assumptions were already being made about the future, ahead of any scientific justification. It was as some sort of palliative to this objection that the initials JET in Community documents were changed to stand for Joint European 'Torus' — as against 'Tokamak' as had originally been the case.

Sadly, Luc died suddenly in March 1972 and into his shoes stepped Lorenzo Enriques, a bright and vigorous physicist from Frascati. Palumbo briefed him on the background to the project and set out his aims for a collaborative European effort. Following this, Enriques made a round tour of all the European laboratories to organise theoretical analyses of different aspects. At one time there were of the order of 100 people working on the study and from whom he was receiving reports. While he and the Working Group were assembling all this data, the Directors were also concerned with formulating a programme that would take into account the enlarged Community, bringing in the considerable effort of Culham, and the more modest programme of Risø in Denmark. By mid 1972 there was no longer any doubt that the Community would be enlarged from January 1973 (although the surprise of the Norwegian referendum in September was yet to come), and apart from the provisions for 1973, the next round of negotiations relating to the projects claiming preferential support over the years 1974-75 would need to take the new membership into account. Moreover Sweden and Switzerland had intimated that they would like to be associated with the Communities' fusion work. It was also necessary to think of the status a future JET might have and how it related to the overall programme. For the present it was concluded by the Directors that work performed within the Associations for JET qualified for preferential support, and that serious consideration should be given to funding some specific contractual work to the tune of 100%.

By this time money from Euratom in general and preferential support in particular was exercising a potent influence on European programmes. In the UK a panel under the chairmanship of Sir Harrie Massey was completing a new independent review of the Authority's fusion programme. Its conclusions, accepted by the Government in March 1973, were that one could now be confident that controlled thermonuclear reactors could be built and it was therefore appropriate that spending on fusion should be increased. It recommended the building of a large experimental tokamak assembly at Culham and the pursuance of the Authority's programme as part of a collaborative European programme. Culham was thus able to go ahead with two tokamak projects led by Gibson: the building of DITE, a Tokamak Experiment to test both plasma clean up by a Divertor technique and Injection methods, and continuation of the design of the Culham Large Tokamak (CULT or more generally CLT) no longer tied to superconducting techniques. There was however, little possibility of CLT being funded without a major input from the Communities. Construction of the French TFR was going extremely well under Rebut's drive and he had started on the design of a large successor, but this too was unlikely to be built without a substantial Euratom input.

The two projects lent urgency to the problem of deciding what should be done once Enriques had completed his report. Although people had been anxious at the beginning to emphasize that this was no more than a collective view of what the alternatives were, it was already clear by November 1972 when Enriques gave an interim report to the Committee of Directors and then the Groupe de Liaison, that the Working Group would be recommending the construction of a really big tokamak and the start of detailed design work as soon as possible. This clearly could not go on under the present committee regime and the Communities would have to devise a new operational structure. Palumbo urged that consideration be given at once to the form this should take, and at the same time to the criteria to be applied to choosing a site on which the machine would be built, the industrial participation and so on. Palumbo saw the movement as a great opportunity, realising that if they did nothing together he would be under pressure to finance two similar machines — one in France and one in the UK — and an equivalent machine at Garching as well — none of which was much more ambitious than the American PLT. Some of the Directors saw his concern as premature. For the construction phase, should there be one, there was time to think, and it might be sufficient for the design phase to pursue a common effort within one of the Associated Laboratories without establishing a new management structure — Pease had already offered as a nucleus the Culham group of six people that would be working on the CLT in 1973 — provided this did not pre-empt a decision on the final site location.

Such was the state of the discussions when Enriques presented his report in March 1973. It recommended unequivocally that a big tokamak of conventional technology (not superconducting) be designed and constructed as a joint European venture in order to explore plasma behaviour under conditions close to those expected in a power reactor. A plasma current of 3 MA was proposed and three different approaches to achieving this were outlined. Broadly these ranged from a high field ($B_T = 10$ T) low aperture, modest major diameter structure to one of low field ($B_T = 3.6$ T) big aperture and big diameter. The Working Group preferred the middle way. In identifying as one of the objectives the study of alpha-particle heating (helium nuclei which are the main product of fusion) the capability of using tritium was foreseen. Significant additional heating from external sources would also be necessary. The proposal, it was claimed, should be treated as urgent if Europe was to make an appropriate contribution to world development in this field. A design group to continue the study in depth should be set up forthwith with a view to the machine being in operation at the beginning of 1978. The building of specialised experiments in the different laboratories in support was also recommended. Cost of the machine was estimated to be around 25-30 MUC excluding buildings and staff salaries.

The Groupe de Liaison meeting which received the report was the first to be held following the enlargement of the Communities and so the first to include British delegates as full members competing for Euratom funding. They had brought with them the plans for DITE and also their design for CLT. This was not only an affirmation of their confidence in tokamaks, but also a reminder of the alternative national proposals in preparation.

The consensus view of the European fusion community, or at least the tokamak community, including the British, was in favour of the collaborative venture proposed by Enriques, and this was given formal blessing by the Advisory Group on Tokamaks of the Groupe de Liaison on 3/4 May. Enriques and his Group had through questionnaires, workshops, analysis groups and personal contacts made a wide consultation at the working level and the result was a proposal more ambitious than either of the two national projections, uncompromising in its aims and scope, and firmly collective in character. Although fully aware of what to expect, the laboratory heads received it with mixed emotion. They doubted, for example, that costs would be much under 50MUC; Pease would have been happier with a simple recommendation to back the CLT; Trocheris (the Director of fusion at Fontenay), already anxious about the start-up of TFR and Rebut's impetuousness, was aware of the private view of many of the physicists at the CEA that their own proposed follow-up was already a foolishly big step. Schlüter was seeing, as he feared, the band-wagon beginning to roll to the disadvantage of stellarators and, because of it, IPP's contention that their big stellarator project Wendelstein VII should receive equal consideration in a European context obtaining little support. Also in a diffuse way the laboratory heads recognised that should they launch into such a project, fusion would never be the same again.

Despite the feelings of trepidation, there was no turning back. Politically, the time seemed ripe for a new Community venture that would involve the new members (Britain, Denmark and Ireland), elsewhere in the world there was talk of big tokamak projects and this was a size that was still just within bounds, but sufficiently important to justify a joint effort — a thesis that appealed, for example, to Braams, convinced by the argument that international projects should be restricted to those too big to be handled comfortably in a national context. Moreover a decision to press on with a design was not yet a decision to build. Accordingly the Groupe de Liaison from its meeting in May issued a formal recommendation to the Commission that steps should be taken to define 'a large torus facility of the tokamak type (Joint European Torus) . . . of about 3MA. The design, construction and operation should be made as a joint European project (and) all arrangements should be such as to guarantee this collaboration'. Pease's offer of Culham for the design centre was accepted and the Commission was informed that a design team was being set up, while the Associated Laboratories were asked to state how they would participate in order that the *modus operandi* could be formally established. The Communities were invited to endorse this action and include the design activity in the current 1971-75 programme.

ORGANISATIONAL STUDIES

The Committee of Directors, plus Horowitz (by then responsible for fusion amongst other research areas within the CEA) and François Waelbroeck (Chairman of the Advisory Group on Tokamaks) met in Culham on 10/11 May 1973 with JET as the main (but not only) item on the agenda. They had before them a working document prepared by Palumbo that foresaw the creation of a permanent Executive Committee with powers like the Steering Committees that managed the contracts of Association. These consisted of representatives of the laboratory concerned and Euratom. Advising the Executive Committee would be a Scientific and Technical Committee, and finally the head of project. As an expedient during 1973 while more formal solutions to the legal, administrative and financial management problems were being worked out, an *ad hoc* Supervisory Board would be formed by a few representatives of the Groupe de Liaison. The various laboratories providing experts would be asked to support their own people during 1973 while Euratom paid their travelling expenses, and the host laboratory—the team would need to be in one place—provided general services and facilities. For 1974/75 the Commission would be asked to include the activity within the preferential support recommendation that had to be sent for Council's approval before the end of 1974.

The basic proposal of establishing an international design team was accepted by the Directors and before the end of the meeting the names of a number of its members had been agreed. Rebut was nominated as the possible leader. Five senior men from the different laboratories were also designated to form an *ad hoc* Working Group under the chairmanship of Rebut which would meet at Culham over a week to draw up a plan for the design phase and the immediate programme of the design group.

Discussions had already been started between the three biggest laboratories on the structure of the organisation within which JET could be designed and built, initially between Willson and Ernst-Joachim Meusel, head of administration at IPP and then with Horowitz for the CEA. Palumbo also was pressing the urgency of defining an administrative committee that could supervise the follow-up to the JETWG's report and which would work out the financial control, the definition of site criteria and the status of staff. There was a large measure of agreement between the major parties on the need to keep a close link with the Groupe de Liaison and everyone assumed that JET when it came to building would be sited at one of the big laboratories. Although anxious not to give the impression that they were forming a 'cartel', these laboratories felt it was up to them to define the terms. They were encouraged in this approach by Gunter Schuster, the head of the Communities' General Directorate for Research, Training and Education, which encompassed fusion. His preference was to confer quietly at first with the big countries before opening a general debate.

So it was that at the end of May 1973 there was a meeting in Saclay of the representatives of the CEA, Culham and IPP. They had before them a long memorandum from Schuster setting out the type of question that would need to be answered before a proposal could be submitted by the Commission to the Council of Ministers. Among the recommendations that were made was that a new 5-year programme for fusion should be prepared that would start on 1 January 1975. This would then be updated each year so that a smooth transition from the design to the construction phase could be engineered. Subsequently the project could be run without fear of a sudden stoppage due to budgetary or other difficulties within the Communities. He considered it necessary to submit a site proposal as soon as possible and urged the Associations who would make up the partnership to arrive quickly at an agreement on the criteria to be applied. He suggested that the Committee of Directors should define the form of site questionnaire to be sent out by the Commission and also establish the rules for evaluating the answers. However, the actual assessment should be made by a Site Committee consisting of nominees of the partners—one per partner—who were of high professional and personal integrity but not personally involved in fusion research, under a chairman nominated by the Commission. A time-table was foreseen that would lead to that Committee presenting its findings by July 1974 so that the Commission could make a proposal to Council in September 1974.

On the basis of 30 scientists and 30 support staff, some 7 MUC over 2½ years would be required for funding the design phase, 50% of which would be contributed by the partners and 50% by the Commission. The three laboratories had already made known their opinion that it would not be necessary to establish any new legal personality for this phase so that a Contract of Association which defined the roles of the partners and the Commission could be set up relatively easily. However, it was desirable to settle quickly the general outline of the structure that would be adopted for the following phases, in order to ensure compatibility. The legal structure as such could be left until the site was known.

Meusel had circulated a long list of questions that needed to be answered in regard to the authority that would be exercised at different levels of management and the means by which decisions would be taken. He pointed out that not all partners

in the Associations had the same type of constitution. Palumbo meanwhile had set down his own views on the form of the future JET structure. He foresaw four possibilities:
(a) integration into an existing Association with supervision by an international steering committee;
(b) setting up a common enterprise under Article 45 of the Euratom treaty;
(c) establishing an enterprise through agreement between states or
(d) by agreement under national laws.

He made no secret of his preference for the first which came to be known as the Plan Palumbo.

It had become accepted that the project would comprise three phases — design, construction, operation — and although nominally design need not necessarily be followed by construction, there was no doubt in everyone's mind that provided a valid design could be produced and governments could be persuaded to put up the money, construction would follow. Implicit also was the understanding that the money needed would be 'extra' money. The Communities throughout the seventies worked on the basis of programmes being agreed individually, to be fought for without reference to any pre-determined ceiling.

At the Saclay meeting, the three partners made clear their main preoccupations. For example, the British were against the creation of another international organisation, preferring a form of Association Contract, but could accept a company established under the national law of the host country — a structure much favoured in Germany. The Groupe de Liaison, on the other hand, had been drawn towards a joint undertaking in preference to an Association which some members felt would be too dominated by the host laboratory. France was more concerned with independence and efficiency — a recurring theme of Horowitz in the years to come. He insisted on the necessity for the organisation to be able to act in its own right using its own name. He explicitly rejected the Dragon* arrangement where the UK Atomic Energy Authority acted as its legal personality and passed contracts on its behalf. This may seem surprising in view of the fact that all decisions regarding Dragon were taken by the project itself without ambiguity and that throughout its existence no doubts were ever cast on the system either from the point of view of efficiency or integrity. However, it must be remembered that the partners in that organisation were the UKAEA, five small countries and Euratom representing the interests of initially the 'Six' and after 1973, the 'Nine', leading to a very confused situation.

Horowitz wished above all to put the project in a situation where decisions would not need to be referred back to the Council of Ministers but would be taken by an independent management. He was totally opposed to the project being managed

---

*OECD Project for the development of the High Temperature Reactor established at Winfrith, England, in April 1959 and terminated in March 1976.

from Brussels on the grounds that the Commission and its servants were simply not competent to do this. He was also anxious that the project should be guarded against one of the smaller countries arbitrarily bringing it to a stop or it being halted for other extraneous reasons such as a crisis within Euratom.

Willson had foreseen a structure in which a Management Board of about seven people responsible to the Commission and national authorities ran the project through a project manager, but it was accepted that the composition should depend on who was financing it and where it was located. All were against Palumbo's idea of creating an 'administrative committee' that included representatives from outside the field, such as CERN, ILL, ESRO/ELDO. They were strongly of the opinion that the Groupe de Liaison and Committee of Directors should be the controlling bodies firmly under the influence of the Associations. If, however, Schuster insisted on a new committee, it should be small and informal consisting perhaps of one representative from France, Germany, Italy, the UK and Euratom and one from the remaining partners.

Meusel made one more appeal for the construction of Wendelstein VII to be considered on the same basis as JET but the other two argued that the whole project was too far advanced for it then to be made European. For a project to be European in the JET sense, they reasoned it needed to be internationally conceived from the beginning. This would not prevent the operation of WVII being 'Europeanised' or for other laboratories to send staff.

With Schuster's encouragement, Horowitz, Meusel and Willson continued their discussions at further meetings and in the middle of August 1973 produced a joint 'Statement of Opinion' that they presented to Euratom.

This endorsed Schuster's original recommendation that the Partners were to be the principal arbiters, consulted before the Commission transmitted to Council any information on JET such as statement of objectives, costs and staffing. It was too precipitate to try and present a new Communities' 5-year fusion programme starting 1 January 1975 and they proposed following the established cycle, but begin at once preparing a 5-year programme for the years 1976 onwards so that all was ready in good time for the decision of the Ministers. This should then turn into a rolling 5-year programme up-dated every year. Moreover, to avoid JET becoming a political football to be kicked around annually, a system similar to that adopted at CERN should be introduced. There each year, the budget for the following year is determined, that for the year after is provisionally settled and a good indication is given of the resources that would be available for the year after — all at fixed prices with a largely automatic process for indexing. (The system, in practice, is not followed to the letter, but no country can apply a veto and departures from the forecast pattern are relatively small. On no occasion to date has CERN entered a new financial year without a budget being agreed for that year).

The choice of site was regarded as of prime importance, but before establishing a formal procedure it was recommended that unofficial contacts be made with national authorities to determine on which sites first, a sufficient supply of energy could be made available, and second, tritium and other radioactive materials could be handled safely. The question of power supplies was far from trivial. During a machine pulse, lasting several seconds, JET would be consuming electricity at the average rate of a city of half a million people. If this were taken as an instantaneous load, the grid feeding it would have to be near by and of high capacity. The load could be evened out to some extent by storing energy in motor generators fitted with massive flywheels, but these were likely to be at the limit of present day experience (in terms of size) and could add materially to the cost. The desirability of having a site where there was already experience of radiation safety problems and which had an appropriate infrastructure to deal with them was evident.

These were by no means the only criteria that would be applied to the site. Other technical criteria included the availability of workshops and computers for example, whereas on the social side were identified the role that national trade unions might play, the cost of living, schooling for an international community, housing and so on.

Schuster's suggestions on site assessment were closely followed and a time-table set down that foresaw the preparation of the site questionnaire and the constitution and programme of the site committee being settled in February 1974, inspections during April and May, evaluation of the results June to July, followed by the recommendation of the Commission to the Council of Ministers in September 1974 and their decision in November. As to structure, for the construction and operational phases, it was hoped that all the currently Associated Laboratories would become contracting parties in an organisation that might be:
– a joint venture according to Article 45;
– a legal personality constituted according to the law of the host country;
– a management contract with the site laboratory.

The main disadvantage of the first was stated to be the need for the Council of Ministers to pass a resolution, in addition to an expression of intent from the Member States, and this would lead to an awkward and inflexible solution. No comment was made on the disadvantages of either of the other two solutions and one might infer that Horowitz was already making clear his preference for the second, modelled on the lines of ILL, Grenoble (which it should be noted was set up with two partners only — France and Germany — although the UK joined later), while Willson was reflecting Culham's preference for the third — assuming a Culham location. IPP would have gone along with either type of organisation.

At this distance it seems strange that whilst there should have been such nervousness about the Council having to approve the structure — and this was not yet a question of the Commission being involved in management — no doubts were raised about the problems of choosing a site. Delegates had in front of them the example of CERN which despite its democratic structure and powerful lobby had failed over 2½ years to arrive at a consensus on the site for its '300 GeV proton synchrotron' from amongst those offered. The new site never was agreed and the machine could only finally be built as an extension to the existing facilities. Admittedly the estimated cost of JET reckoned at 40-60 MUC was small in comparison with CERN's demand of 2000 M S Fr. so the emotions raised might be more moderate, but the issues remained the same. The European Southern Observatory and the European Molecular Biology Laboratory are both quite modest operations and quite removed from Community politics, yet their final location was decided only after long and protracted negotiation (and large financial inducements). Site selection, like everything else involving a choice between countries, is inevitably a political question and the apparent naivety with which the site for JET was approached was a commentary on the sheltered environment (in the context of international politics) that fusion had enjoyed up to then. At the same time it reflected the keeness of the fusion people to get on and not be put off by the difficulties that would certainly arise.

Very quickly the framework of the project beyond the Design Phase won general approval. JET was to be an extension of the existing Associations with a scientific programme drawn up by these Associations and integrated into the overall programme of the Communities. Implementation was to be entrusted to a strong Board of Management of perhaps eight people drawn from Euratom and the Associations with full powers to take decisions within the overall budget. The Board would report to the Groupe de Liaison and the Committee of Directors. The size of the project was still not clearly defined as the figure of 50 MUC was felt to be probably a little optimistic, but in round terms, JET could be thought of as costing up to 12 MUC/a during the construction phase and some 5 MUC/a during the operational phase — the majority of which would come out of Euratom budgets (that were in general contributed by national departments remote from the fusion sector). For comparison, the average annual fusion budgets in Europe during 1971-75 totalled 58 MUC to which Euratom was contributing at the rate of about 14 MUC/a *i.e.* roughly a quarter. Assuming that Euratom would pick up some 60-70% of the bill, the cost of JET to the individual Associations would be quite modest (~7% of existing budgets). Indeed, both Braams and Vandenplas were moved to question the advisability of making the Euratom contribution any higher than for priority programmes, suggesting that a general increase in support for fusion would be more appropriate. Altogether, JET was an attractive concept: too big for any of the individual Associations yet offering no serious competition to existing programmes.

Design Phase Agreement

Against this broad background of agreement over the shape of the later phases, the Committee of Directors got down to the immediate task of defining exactly the responsibilities of the different bodies involved in the Design Phase. Until this was done no funding would be available from the Commission and to fill the gap,

Culnam agreed to cover design contracts and put 100,000 UC at the disposal of the Design Team. By the end of 1973, the terms of a Design Phase Agreement had been largely decided and early the following year it was possible to complete the final drafting. The Agreement was deemed to have come into force on 1 October 1973. Signatories, in addition to Euratom were the CEA from France, CNEN and CNR from Italy, the Danish Energy Agency, Belgium, FOM from the Netherlands, IPP and KFA Jülich from the Federal Republic of Germany and the AEA from the UK.

The Agreement formalised the status of the Supervisory Board that had been acting up to then and its two subsidiary committees — the Committee for Scientific and Technical Matters and the Committee for Administrative Matters. The Design Team was to be drawn from the Partners, its members fairly shared according to the different Associations' budgets. Salaries would be paid from the home base while the Commission looked after allowances and travel expenses. The AEA was to make available suitable accommodation at Culham and would see that services were provided, for which they would be reimbursed by the Commission to the tune of 75.85%. Explicit mention was made of the work to be carried out under the supervision of the Head of Project by the Associations, a guide-line figure of 2.5% of their budgets being foreseen for 1974 and 1975 for which the Commission would reimburse an additional 20% up to a limit of 0.6 MUC. Procedures for putting out work to third parties was also specified, the amount cited being 1 MUC initially and a further 2 MUC for long delivery items once the basic design report had been approved. The detailed design report was to be ready by June 1975 at the latest.

It is striking that there was total acceptance of the basic principle that the project should be pursued as a project of the Communities. Even to the recently joined British, it was entirely natural to see Euratom as the central and essential coordinator of a common enterprise in fusion. Euratom money was, of course, a major inducement, but this in itself was a testimony to the efficiency of the administration and its negligible cost (less than 1% of the turn-over). The absence of any discussion over the legitimacy of the sponsoring body constituted a silent tribute to the painstaking work of Palumbo and his staff over the previous decade and the sense of community induced by the Groupe de Liaison and the Committee of Directors.

DESIGN PHASE OPENS

By the time even the Statement of Opinion had been drawn up, and certainly before any of the details of an Agreement had been worked out, the Design Phase had, in effect, already begun with an élan that seems now quite dazzling. Against the background of the successful operation of the TFR at Fontenay, where stable pulses of 200kA lasting for ¼ second and temperatures of up to 2.5 million degrees had been recorded, the Committee of Directors had met on 15 June 1973, first in special and then later in ordinary session, to decide the next steps. Having heard Rebut's report from the *ad hoc* Working Group which sketched out the broad lines of how the design would be pursued, the manpower and costs needed, and the support to be expected from the Associated Laboratories, the Committee in quick succession:

1. Named the members of the Supervisory Board which would oversee the Design Phase and report back to the Committee of Directors.
2. Designated Romano Toschi as its Chairman.
3. Chose Culham as the site where the Design Team would be located (without prejudice to the final choice of site for the construction).
4. Invited Rebut to head the Design Team (so preserving the principle that the project leader should not be a member of the host laboratory), and nominated as senior staff Roberto Andreani from CNEN (in the event replaced by Enzo Bertolini), Dieter Eckhartt from IPP, Jean-Pierre Poffé, a Euratom physicist at Fontenay, and Gibson and David Smart from Culham.
5. Gave undertakings regarding the detachment of about 25 scientific and engineering staff. It also established the principle that JET should not take on its own personnel but should be staffed by people seconded from the Associates. The terms under which these should be engaged were subsequently covered by a simple amendment to the mobility agreement between Euratom and the Associations that extended the existing agreement to the Danish Atomic Energy Commission and the UKAEA and provided for the payment by Euratom of foreign station allowances according to the agreed scale direct to the staff seconded to the JET Design Team.

6. Explored in detail the role and responsibilities of the Management Board that would run the project during the construction phase.

The Supervisory Board (SB) met for the first time on 6 July 1973, interpreting its mandate as being 'more responsible to the nine Associations than to any other body'. It ruled against the Tokamak Advisory Board being given responsibility for technical supervision although at its next meeting designated it as the permanent scientific committee for JET. The SB sketched out its general method of working and set a level of £10,000 for contracted work that would need its authorisation. On the basis of the estimate discussed in the Committee of Directors, it decided that the Commission be asked to provide 8.5 MUC for the years 1974-75 being 10% of the estimated capital cost plus salaries, and the different Associations be invited to organize their work so that they would be contributing about 3.5% of their budgets in terms of staff and studies directly related to JET.

The Design Team

Rebut came to Culham in September 1973 and by the end of the year an international team of 30 people had been assembled. It is a popular image that he arrived with the whole of the design already in his pocket — or at least in his head, but this does less than justice to his immediate colleagues and to the many scientists with whom the design was discussed over the years, and it ignores the changes that were made as the design went on and the huge amount of detail that had still to be filled in. Rebut had certainly given most thought to the overall engineering concept and to its technical feasibility and he had delivered a paper to the French Physical Society setting out his own ideas on what a large tokamak should look like. A very rapid thinker with a phenomenal grasp of engineering detail, he had a tendency in those days to betray his Normandy origins, and particularly in committee, give the impression of having already made up his mind on whatever issue was under discussion and of having very little time for any view-point other than his own. He was regarded at Fontenay as something of a 'vieil ours', brilliant but liable to take too much on his own shoulders and break things. The story goes that during the management meeting (from which he was excluded) called to discuss whether to risk taking TFR to full field, Rebut was out there doing it — and it didn't break. In practice, Rebut has always been open to informed argument and once persuaded on the rightness of a course will rapidly analyse the implications for the system as a whole and will support that course from then on.

Before he came to Culham he had formed very clear ideas on priorities and the essential configuration of the design, and even if he went to the extreme of the Enriques proposals, only in the cross-section of the torus did he explicitly run contrary to the recommendations that had been spelled out. Moreover in the Enriques analysis it was recognised that the particular ratio of major to minor axis chosen as 'standard' was only consistent with the middle design and further work was necessary to make both the high and low field concepts self-coherent. In the Enriques analysis high plasma current was given the 'primary figure of merit' and this thesis was certainly adopted as well as the principle of using existing technology (water-cooled copper coils for example). It was in looking always to minimize mechanical stresses while exploiting to the maximum every cubic centimetre, that led Rebut to propose a non-circular cross-section for the toroidal vessel and a quite novel coil mounting structure. These he had already presented to the *ad hoc* Working Group at their session in May and secured their approbation.

Rebut's colleagues on the Project Board were not entirely the carefully balanced team of scientists and engineers, chosen for their complementary skills that he might have selected given real freedom of choice. The Board was essentially a delegate group of senior men from the principal laboratories who could—and would—be transferred at that moment. Although fusion physicists abounded, the number of experienced scientifically trained design engineers in the field was very limited, and some of the Board members were to find themselves in charge of major aspects of the design, having had no previous experience in that sector. It says much for the adaptability of these scientists and the strength of Rebut's leadership that they could so quickly settle down into a harmonious collaboration.

Of them all, only Gibson had been in the JET Working Group and could be regarded as an automatic choice. A fusion physicist of many years standing, he had been responsible for devising the physics programme for CLEO at Culham and recommending the new programme for the laboratory there which led to the construction of DITE. He had played a large part in defining the basic characteristics of CLT as well as serving as Culham's scientific representative within Euratom. He was the principal advocate of the low field machine, whereas Rebut's own leaning at the beginning of the Enriques study was towards an extrapolation of the TFR with higher toroidal field. Although more amiable on the surface and certainly more diplomatic than Rebut, he can be equally stubborn when it comes to either scientific or organisational matters that he considers involve basic principles. Smart, the second man from Culham, the oldest in the group was head of the Engineering Design Division at Culham. Primarily an electrical engineer, his responsibility for the poloidal field system was logical. He had in the *ad hoc* Group promoted the concept of the CLT with a circular cross-section and no iron but had been won over to the concepts of Rebut, for whom he developed a great admiration. Smart knew Culham inside out and was able to advise the Project on how to fit into the existing system. Calm and competent, he was able to smooth the relations with the host laboratory and take advantage of the facilities that were available without compromising the Project's determination to stand apart. In this he was much aided by Roy Bickerton, Vice-Chairman of the Supervisory Board under whose aegis the Design Team was established within Culham. Although Bickerton would seem to be the antithesis of Rebut—studied, reflective, basically pessimistic—he still believes that in fusion you have to push on with empirical experiments. Independent in his thinking and apt to be cutting in his comment, he and Rebut nevertheless formed a firm alliance, enjoying a mutual respect (helped, no doubt, by their common interest in sailing).

Quite new to tokamaks, though not to fusion, was Poffé, a Belgian physicist who became a staff member of Euratom in 1960 expecting to be sent to Italy, but when the placing of fusion work in Ispra was stopped, he joined Fontenay under

Members of the early design team being shown round Culham. From left to right: A. Gibson, P-H. Rebut, G. Venus, M. Huguet, J.-P. Poffé and D. L. Smart.

the Contract of Association and there continued to work mainly on mirror machines. When the mirror programme was shut down he moved to Culham to take responsibility for forward planning and the future site. Solid in frame and in character, he had been a colleague of Rebut over many years and they had no trouble in cementing their association at another place.

The most difficult transition was for Eckhartt, a physicist from IPP who had also spent some time at Princeton, despite the fact that he was amongst the first to arrive at Culham. He had already crossed swords with Rebut in the past and his position was not made easier by the fact that Garching tended to be the least enthusiastic for the Project in general and the most critical of the Associations in regard to the basic design features. He himself was made responsible for certain materials problems and for the vacuum vessel design which was entirely novel to him. Nevertheless, being basically of a conciliatory and loyal nature he was prepared to learn from Rebut who had clear ideas about what he wanted and he was soon able to make his own contributions.

Last to join the Board was Bertolini from Frascati. Originally attracted to fusion when at CERN, he had moved to Frascati and subsequently specialised in magneto-hydrodynamic (MHD) power generation—related to plasma physics but not really fusion. He had been invited to head the Frascati Tokamak team but had preferred to go back to the USA where he had developed regular teaching responsibilities. On returning to Italy he found MHD was being phased out and so moved over to tokamaks and the conceptual design of a reactor. Due to go back to the USA in the Autumn of 1973, he was encouraged to come first to England, where he found himself offered the toroidal power supplies 'because that was the only thing left'. However, he found Rebut's designs close to those he had been working on for a reactor (notably D-shaped toroidal coils) and was stimulated by the prospect of working with Rebut whom he regarded as a human dynamo of ideas. He changed his mind about the USA, joined the team in January 1974 and set about becoming an expert in his allotted field.

One further name should be mentioned at this time; that of Michel Huguet. Although not formally a member of the Project Board he was one of the original group leaders, responsible for the toroidal coil and mechanical structure. From Fontenay too, he was one of the original team working with Rebut on the TFR. They knew each other well and Huguet has the capacity for understanding Rebut's flow of ideas even when they are not too clearly expressed as happens when his mind runs ahead of his tongue. One of Nature's gentlemen, but tough for all that, Huguet was able to cope with Rebut's continuous intervening in all the minutiae of the design and also with the relentless pressure that he exerted. For Rebut too it was important to have people such as Huguet and Poffé with whom he could discuss details in his own language without any of the frustrations associated with communication in a foreign tongue.

The international language of fusion science is certainly English and all the Project Board were 'reading fluent' in the language, but this is not the same as talking, arguing a case, explaining a difficult thesis . . . It is easy to underestimate the strain of changing one's working language and the frustrations associated with having to search for words in the cut and thrust of discussion. Moreover, even if scientific terminology is often international, the vocabulary of engineering detail is huge and much more local in its origins. However, the Design Team took this in its stride and learned to cope even with the local Berkshire accent.

Harmony within the Project Board was possibly helped by the formal collegiate nature of the Board's structure as defined by the Committee of Directors. Members were in a real sense representative of their laboratories and they were part of a decision making body whose collective conclusions were communicated by the Head to the Supervisory Board which, in the event of disagreement, would be the arbitrator. Part of the argument for such an arrangement was no doubt to ensure that Rebut could be contained, but the main reason was that at the level of the Committee of Directors, the basic concept was of a corporate effort responsive to the wishes of the parent laboratories. However, in the subsequent struggles with the Supervisory Board, for example to get agreement on upgrading the machine, it is evident that Rebut and his colleagues were united, and they were together in urging the bold approach, even if this meant opposing their own people.

Basic Design

Until then (and not excluding TFR which was the biggest in the world) toroidal machines had looked rather like overgrown inner tubes from a tractor wheel wrapped round with heavy cables, some following the major circumference, others the minor. Various pipes were connected to the torus, some of which were used to pump out the gas inside, some to give access to measuring instruments — the diagnostics in fusion jargon — while others would let in hydrogen gas as needed, give access to the inner shell of the vessel and to devices inside designed to limit the size of the plasma and so on. The whole looked rather complicated and perhaps a bit messy in the way of experimental machines where temporary equipment tends to be hung on in an apparently haphazard way, but there was room to move and a complete machine would typically be housed in a room say 10m × 10m by 5m high.

One feature they had in common was a rather massive framework, disproportionately heavy for a vessel that contained only a rarefied gas or nothing at all. The frame is needed to withstand the very large forces that can be generated when conductors carrying heavy currents are in big magnetic fields — the same forces that drive an electric motor. One of the basic rules of electricity says that if a current is flowing in a conductor at right angles to a magnetic field, then a force will be exerted at right angles to them both. Only when the conductor and the field are parallel is there no force. In a tokamak, the objective is to produce a twisted field by combining the circular (poloidal) field produced by the current flowing round the plasma ring with a field parallel to the current (toroidal) produced by coils encircling the minor circumference. Inevitably therefore the conductors are subjected to twisting forces. Moreover strict parallelism is only seen in straight conductors so that even a single coil carrying current is subject to forces trying to stretch it. Place a number of these round in a ring and they interfere with each other, pushing in towards the centre. The size of this push from each of the toroidal coils in a machine as big as JET can be over a thousand tons.

Rebut's design, as was explained in the team's first report, EUR-JET-R1, produced in November 1973, was based on three basic principles, over and above the primary requirement of a high plasma current (and so large poloidal field) that could be held constant over many seconds. They were:
1. reduce every stress to the lowest reasonable value;
2. provide for the possibility of remote replacement of a component come the day when the machine would be radioactive;
3. be as compact as possible to get maximum value for money.

The first principle led to the rounded D-shaped form for the vessel and toroidal field coils — as this was the natural shape the latter would take up; the second made it essential to put the poloidal field coils outside the toroidal (so the outer ones could be lifted (or dropped) out of the way and a sector of chamber then pulled out with the associated toroidal field coils round it); the third resulted in a daring conception whereby individual casings for the toroidal coils were omitted and instead these coils were incorporated into the support structure and jammed tight together in the middle, the centring force being transmitted through the inner poloidal field coils to a central column which formed the common inside leg of a transformer with eight symmetrical rectangular yokes. To maximise on every centimetre of space, the centre column was even waisted in the middle. The poloidal field coils, it should be noted, fulfil two functions: the inner, mainly form the primary winding of the transformer (of which the plasma is the secondary); the outer, shape and position the plasma inside the torus cross-section.

A highly symmetric field around the torus is required and this was achieved by having 32 equally spaced toroidal field coils surrounding the all metal vacuum vessel which was made up out of eight similar modules. The modules could be pre-assembled and then slid into position between the outer transformer yoke sections and welded to the adjacent sectors. For the torus the mean major diameter was set at 2.82 m and the D section 2.38 m across × 3.96 m high. Overall, the bare machine would measure 14 m across by a towering 10 m high.

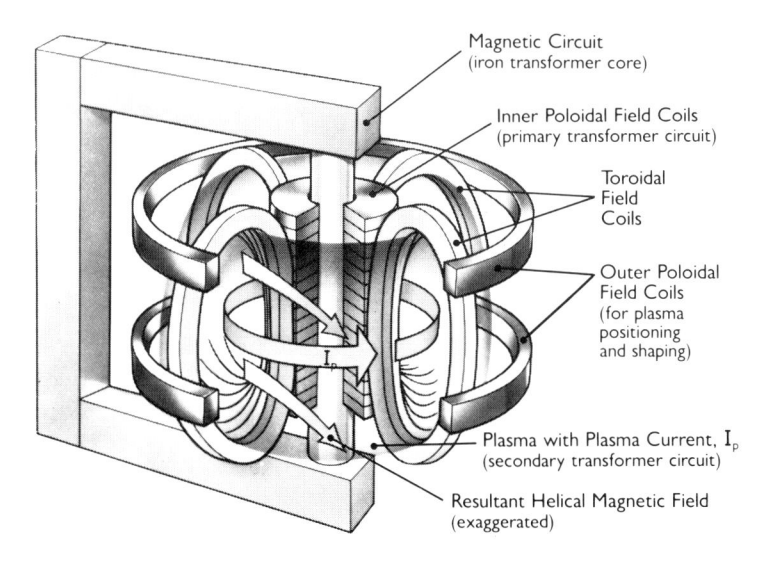

Simplified schematic drawing of the JET tokamak showing its magnetic field configuration.

JET's Aims

In that report a hint of what was to come was contained in the phrase 'it is unlikely that ignition will be achieved with as small a value of plasma current as 3 MA' and a sketch of the poloidal field pattern to be expected was plotted for a plasma current of 4 MA. Rebut had for long held the view that the 3 MA which had been chosen as the minimum value necessary for confining a high proportion of the alpha-particles produced in fusion would not be sufficient to reach ignition.

Although the Supervisory Board (SB) was meeting at almost monthly intervals, at its third meeting in October 1973 there were already protests that the Design Team was placing them 'in a situation of quasi-accomplished fact'. This was also provoked by the initiatives taken by the Team who wanted to put out design studies to manufacturers of their own choosing for certain key items like power supplies, prototype coils, and tritium handling, whereas the majority of the SB wanted a formal tendering procedure. This was subsequently worked out by the Committee of Directors, and presented at a joint meeting with the SB and the Project Board at the end of November. It provided for JET to be presented by the Associations with a list of relevant companies and for a jury to assess offers for contracts worth more than 0.5 MUC. At the same time the principle was established that the awarding of a study contract gave no priority when it came to the manufacturing stage—implying that the results of study contracts would be fully open. Not all the Associations saw the necessity for going to industry at all, but Rebut (supported notably by François Prevot, the representative of Fontenay) insisted that it was vital to benefit from the experience of industry and involve them at an early stage in the manufacturing problems of the large and novel components that would be ordered.

By the beginning of 1974 the Committee of Directors had laid down rules for the SB to follow and set up an Administrative Committee and a Scientific Committee, the one to act a posteriori particularly in regard to contracts, the other to offer independent advice on the machine and its objectives. The legal basis for the whole operation was covered by the Design Phase Agreement between Euratom and the nine Associations valid until 31 December 1975. Conscious of the smallness of his group and the difficulties he was already encountering of getting the Associations to co-operate fully in undertaking design studies, Rebut was irked by all the superstructure and urged the SB to try to ensure only 'constructive interference' in what his team was trying to do. He had expressed the same view to the Groupe de Liaison when it discussed in December 1973 the final draft of the Agreement.

At the Directors' level however, there was considerable concern about the future for fusion in the light of the oil crisis that broke upon the world towards the end of 1973 when virtually overnight the price of crude oil quadrupled. Far from encouraging investment in a long-term energy source that had yet to be demonstrated, they feared it would cause funds to be diverted to more immediate short-term aims. In December, the Council of the Communities had approved a fusion budget for 1974 and 1975 of only 14.9 MUC instead of the 17.6 MUC proposed, of which 0.7 MUC was a real cut in the part of the programme covering preferential support, and the other 2 MUC concerned that part of the JET budget that would cover the long term orders in preparation for the construction phase. Despite there being for the 1971-73 period an unspent sum of about 2 MUC, the Group for Atomic Questions (the relevant (political) body advising COREPER, the Committee of Permanent National Representatives in Brussels which prepares the dossiers for the Council), had not been willing to accept this provision immediately and had in effect put off the decision on 1975 pending submission in September of the proposal for the next 5-year programme (1976-1980). In its first brush with the political sector doubts had been raised about JET's longer term funding which was not a good augury and there was real anxiety about the level of expenditure that the Member States of the Communities would be willing to support.

It was understandable therefore that there was a cool reception to Rebut's proposal made to the Supervisory Board at the end of February 1974 that the plasma current should be increased by a factor of two and the toroidal field raised by 40% to correspond. This would at the same time require another 100 MW of driving power but could be introduced as a two-step approach, first to 3 MA and then 6 MA. All this was to be achieved over 18 months operation after which experiments would start with tritium. Gibson followed this with the proposal that as the machine would then have a very limited life, provision should be made for building a 'second load' alongside, i.e. another tokamak making use of the power sources and services installed for the first.

These ideas, apart from creating concern over the size of the step that was now being proposed and the financial implications, threw into relief the need to be clear about what JET was meant to produce. If it was simply a step in size, that was one thing, if it was to achieve ignition and therefore establish the elementary basis for a reactor, that might be another. To appreciate the difference a somewhat deeper appreciation is needed of the state of fusion science at that time.

In its global concept, a tokamak as we have already seen comprises a means of producing and confining a hot toroidal plasma by magnetic fields created in part by external coils and in part by the induced current in the plasma. The fields can be pictured as combining into a series of concentric elastic stockings that wind round the torus twisting as they go. Individual field lines form helices round the minor and major circumferences. An individual electron on its own would follow one of these helices continually rotating round it, and making about 2½ revolutions of the main to every one of the minor circumference and circling the torus say 5000 times a second. Its more ponderous companion the ion (in the ordinary hydrogen atom the nucleus is some 2000 times the mass of the electron and in deuterium and tritium, respectively twice and three times as heavy again) would be doing the same thing in the opposite direction but much less quickly. Now if we think of a real plasma, at a temperature of tens of millions of degrees, the particles are in a great state of agitation and even though the density may be only one millionth that of the atmosphere, in addition to circling the torus they continually bump into each other. The deflexions that result provoke great irregularities in their trajectories. The collisions serve to distribute the energy and occasionally when two nuclei collide head on, fusion will take place. Nuclei must however expect to make many thousands of turns of the torus before fusing to another. All the moving particles create electromagnetic fields so they are liable to become involved in some local effect they themselves have produced. As in a rush hour crowd, the individuals have their own frantic motion peculiar to themselves, yet participate in collective movements of small and large scale some of which in a plasma is seen as current and some as instabilities that can cause the plasma to crash into the surrounding walls. The plasma physicist is thus trying to learn individual behaviourism, crowd psychology and traffic control all at the same time.

So complex are the motions, recourse has to be made to describing them by empiric rules that are based on experience and only to a lesser extent on the science of individual reactions. But whilst one might succeed in making convincing models of the plasma behaviour in one device, extrapolating this to a second device with different characteristics — even those as simple as torus dimensions — requires an act of faith. There is, however, little alternative and throughout the evolution of fusion, the so-called scaling laws that have been devised have been subject to continuous revision. Inevitably at any given moment there is a temptation to regard the latest versions as 'true', and while virtually all physicists on their own are conscious of their tentative nature, they too are subject to collective phenomena and as a group are capable of generating a confidence in the extrapolations that outsiders might regard as simply wishful thinking. That said, it has also to be acknowledged that without empiric laws, and intuitive convictions, fusion would have made little headway. These are often the only guidelines to determining the direction in which further research should proceed.

The Enriques design current of 3 MA had been chosen on the firm grounds that the poloidal magnetic field so created would keep in the majority of alpha particles produced in fusion. A high current had also been promoted in the belief that the product of density and confinement time (the Lawson number) went up with current and might even be proportional to current squared. Confinement time should not be confused with plasma duration time. Confinement concerns the energy confinement and the efficiency of insulation. While expressed in units of time it relates the energy held to the rate of energy loss. An analogy can be made to a bath with a leaking plug into which water runs. The level will build up until the outflow equals the inflow, from which can be deduced the average confinement time in the bath of the water coming in. This is quite independent of the time the water is left to run.

Starting from the precept that the current was to be maximised, the aspect ratio of the machine (the ratio of major to minor radius of the torus) had been chosen to give minimum cost consistent with a gas density corresponding to existing experiments and a value q, the ratio of toroidal to poloidal field strengths sufficiently high (up to 3) to prevent gross instabilities disrupting the plasma. Overall cost is closely related to ampere terms in the toroidal field coils and minimising this led to the choice of a tight chunky torus. Considerations of maximum field strength that could safely be generated and stresses that could be tolerated in the coils suggested an aspect ratio of something less than 2.5. However, so the argument ran, there was no virtue in having a circular plasma and torus cross-section. A vertical elongation giving a squashed ellipse or D-shaped form to the toroidal coils and the vacuum vessel had the double merit of minimising bending moments while nearly doubling the volume hence current potential. It was also in line with Rebut's inner conviction that ideally a fusion device should look like a sphere and while one could not avoid putting things in the middle, one should approach the ideal as closely as possible. The net result of this approach was a tokamak of quite a new shape to the European community with a very large volume, but an average toroidal field (B_T) at the bottom of the range.

Pumping in more current into a plasma does not necessarily raise its temperature and there was, by then, a strong body of opinion, notably in the USA that however high the current was, it would never be sufficient to raise the temperature to the levels needed to produce a significant number of fusions. Much additional heating would thus be required. In contrast to a conventional electric fire (or light bulb) where, if power is available, because the resistance goes up with temperature the material can be made hotter and hotter until it disintegrates, in a plasma the resistance goes down with temperature, so heating becomes less and less efficient. Rebut nevertheless, had still to be convinced that additional heating was essential and was keen to raise the plasma current in order to improve confinement and raise the temperature, stimulated by the ambition to reach ignition.

But was ignition really the objective? The SB was far from sure and within the ranks there were deep divisions of opinion. Probably the most general reason for hesitation was nervousness over setting a goal which might not be attained, so inviting adverse criticism and putting in jeopardy future funding. Opinions differed, however, on how big were the risks of overreaching present capabilities, and putting too large a fraction of available resources into one experiment. There was also the more scientific argument that the learning process should be progressive and not a succession of discrete islands, countered by the thesis that it was pointless to spend time in areas that were distant from the technological goal, as they might turn out to be irrelevant and even misleading. Whereas Pease was convinced that the essential purpose of JET was to do experiments on a reacting plasma, IPP believed that studying a wide range of plasma conditions was the most important and only when this had been done should one think about tritium reactions.

The arguments were vigorous, but finally agreement was reached on the wording of the aims as set out in the Project's reports. These became incarnate in the ultimate agreement as 'to extend the parameter range ... up to conditions close to those needed in a thermonuclear reactor'. The compromise also gave birth to the notion of a basic and an extended specification which had the double merit of dividing costs and giving flexibility to the development programme.

Rebut's proposal of aiming for a current capability of 6 MA was not rejected outright because in part it corresponded to the recommendations emerging from a workshop that had just been held of representatives from the different laboratories (with Soviet participation) to discuss the JET parameters. At the same time there was concern in the scientific community over the very tight aspect ratio, and the low B_T that had been chosen, these being so different from previous designs. The compromise that was made was that the design had to provide for 3 MA to be achievable in a circular plasma and if all went well and results seemed to justify it, advantage could be taken later of any higher capability. The workshop had also dealt with other aspects of the design such as the control of heavy impurities which radiate much energy even when present in very low concentrations. IPP in particular remained to be convinced that cleanliness and a careful choice of all the materials in the vacuum vessel would be enough. Various approaches to the provision of limiters which define the outer edges of the plasma were discussed and IPP people strongly advocated the inclusion as in ASDEX of a magnetic divertor, a system of bending some of the field lines out of the mainstream so that the outer layer of particles containing the contaminants follows them and is then dumped.

Studies of such a device for JET had been carried out by Peter Noll, from KFA, where there was a programme of research on plasmas of elongated cross-section that had resulted in strong support for Rebut's ideas for JET. Noll was associated with the Project from the beginning, having been a member of the JET Working Group and a Chairman of the Tokamak Advisory Group. In his gentle meticulous way, Noll produced a number of studies but he concluded there was insufficient evidence of either the necessity for a divertor in JET or its likely efficacy. Moreover, all designs took up space in the vacuum chamber and would seriously limit the plasma current. The same for 'adiabatic compression'. This is a technique of forming a hot plasma in the outer part of the chamber, then decreasing its major diameter suddenly which thereby increases its temperature (as when compressing the air in a bicycle pump). IPP were very keen on this as a means of heating the plasma, whilst avoiding the instabilities due to 'trapped particles' that many believed would arise at high temperature, and in the JET Working Group had even proposed an elliptical section chamber with the major axis horizontal. Such a chamber would, however, be heavily stressed as well as wasteful of space and had been estimated to put up the cost of the magnet and power supply by a factor of 2-3 at least. Although Rebut was persuaded to make provision for some compression, he was never really convinced, turning the argument rather into a support for the bigger cross-section.

Workshops were important both scientifically and psychologically in identifying the different Associations with the JET design. A great deal had to be done outside the Design Team to ensure that JET developed as a true community project, but also because the Team's efforts were limited. This was particularly true for the development of additional heating methods, the subject of the second workshop held on 28 February and 1 March 1974 coincidentally with one on technical problems. External heating was vital if the range of operating regimes was to be properly investigated. Two main techniques were under development. The leading method was based on the injection of high energy beams of neutral deuterium which had the added merit of being able to induce some fusion reactions directly (in a tritium plasma) so should the performance of JET be lower than had been hoped, the behaviour of alpha-particles could still be studied. The alternative approach was to use RF waves to raise the ion or electron speed. Both techniques were the object of development programmes in the Associations and their application to JET would be needed soon after JET itself was commissioned.

The Design takes Shape

The first major design report EUR-JET-R2 was produced in April 1974 and was examined by the SB that same month. Palumbo participated in the meeting for the first time and made clear his nervousness at the boldness of the proposals. Both he and Bickerton would have been more content to settle for a higher B_T machine with a 3 MA maximum current, instead of one capable of 6 MA. Gerhardt von Gierke, who as Director in charge of tokamak development at IPP was their representative on the SB, was more preoccupied by the need for a divertor and adiabatic compression, but had still fundamental doubts about rushing to get a reacting plasma and wondered whether tritium should ever be put into JET at all. Rebut, supported by Prevot, countered with the argument that such approaches would lead to a machine that simply reproduced the results, five years later, of the American PLT, the tokamak under construction at Princeton for which approval had been given in the Spring of 1972. Not quite an accurate comment; PLT's maximum design current was 1.6 MA in a torus of 130 cm major radius and 45 cm minor radius, but the basic force of the argument was there. JET had to be bold to be worth doing and the laboratories came to recognise this even if Palumbo continued to advocate a course that was more evolutionary and one that would receive more direct moral support from American plans. It must also be acknowledged that despite the pressures put on the Team to produce an analysis of alternative approaches, the Board never really had an opportunity of examining any, other than in the guise of a justification for the optimisation made. Certainly they were provided with the results of a computer optimisation programme that analysed the consequences of starting from different criteria weightings, but they never had the chance to consider any major departure from the original concepts such as a high density, high B_T approach.

Palumbo had a further reason to be cautious. The first estimate of the cost was in the region of 85 MUC at January 1974 prices of which 49 MUC was for the base machine, 6 MUC for diagnostics and 3 MW of additional heating, 12 MUC for buildings and 18 MUC for construction staff. This was outside the guidelines and would require forceful justification. For Braams, the total cost was the principal worry because of its repercussions on the rest of the fusion programme. Already alarmed by the financial implications of the developments, he had been trying to persuade FOM and the umbrella research organisation ZWO to hand over responsibility for his fusion institute to the wealthier Reactor Centrum Nederland on the grounds tht JET was an energy experiment. He also exploited as argument the IAEA Workshop on Reactor Studies held at Culham early in 1974. FOM and ZWO were in favour of the transfer, as well as RCN, but not the Government which refused to see fusion as yet being mission orientated. Braams having been operating on static budgets for years, had grave doubts that the Netherlands would be willing to join an operation of the size now contemplated.

Despite the concern, Rebut and his team pressed on with their conception of what JET should be. R-2 was modified to take account of the drafting comments and in July, R-4 (a slimmed-down version of R-3, the full cost and manpower analysis without the detailed contract estimates, but including a contingency allowance) was produced as an annexe to R-2. Total hardware cost including 20% for contingency was estimated at 55.2 MUC (3.6 MUC for additional heating and the same figure for diagnostics) to which was added 10.3 MUC for buildings, 22.1 MUC for manpower and 15 MUC for the first year's operation, making a grand total of 102.6 MUC.

That same month the competition between the Oak Ridge National Laboratory and Princeton came to a head; Princeton won and the Americans launched their own big tokamak project to follow PLT.

At the meeting of the Groupe de Liaison held in conjunction with a meeting of the SB in Erice in September 1974, the R-2 design parameters and new cost estimates after much discussion were finally accepted as a basis for continuing design. Von Gierke had withdrawn his objection to the vessel shape in return for an additional expenditure of 2 MUC on power supplies to allow adiabatic expansion. He maintained however that a figure of 104.6 MUC was still marginal and that the sum should be increased to 115 MUC in order to provide a reserve fund that could be used in emergency to off-set inflation or to fund improvements that were perceived to be necessary. His view was coloured by the detailed costing IPP had been doing on ASDEX, their follow-up to Pulsator that had come into operation the previous year. Although not strictly comparable to JET, being half the linear dimensions and designed for one sixth of the current, it had nevertheless shown how overall costs scaled with size. Jülich was of the opinion that building costs had been underestimated and even Toschi who, as Chairman of the SB, generally found` himself defending the Team's views, felt that not enough allowance had been made for the unexpected. The upshot was to bump up the estimate of machine cost to 65 MUC — and then to consider ways of cutting back if subsequently it should seem to be necessary. Von Gierke was in favour of stretching the construction time to five years, Rebut to building a phase 1 machine with just the bare essentials and then adding on later.

Come January 1975, however, and the SB put its foot down. More detailed studies of the central region of the machine had convinced Rebut that the waisted core and angled poloidal coils formed too complex a system with too many unresolved stress problems. The region was therefore re-designed in the form of a cylindrical core and the vertical sides of the toroidal coils on the inside of the D were straightened. This, of course, required other adjustments to the D form and lost space so Rebut was increasing the machine size (height and diameter) to compensate. In terms of hardware this meant 600 ton more copper, 1000 ton more iron in the core and 50 ton more in the structure. It would mean in compensation (and, it was said, for only 1.5 MUC extra) the maximum current could be 6.7 MA. The SB members were not pleased to have this thrown at them without warning and although Prevot would have accepted a price ceiling as the limiting criterion, the majority laid down the rule that no increase in overall machine size could be accepted and whatever problems arose in the future, the dimensions of R-2 were not to be exceeded. Nor were they too happy at the prospect of a consequential reduction in what many already considered to be a too low B_T.

Relations between the SB and the Project Board were rather strained during this period, and to prevent a recurrence of this sort of breakdown in communication, interim reports were from then on required of the project. The SB complained of unilateral decision-taking in regard to design and study contracts, of attention being given to speculative long-term objectives to the neglect of immediate problems, of too little concern about the mechanical aspects of the design, and the absence of fault analysis, while the Project complained of niggling interference, the time spent on paper work and above all, the failure of the Associations to provide sufficient staff. Currently they numbered 44 compared with the 80 or more needed to launch the construction programme. The final irritant to the Project Board was the demand made in March by the Groupe de Liaison on the instigation of Horowitz and with the support of Rudolf Wienecke, the Chief Scientific Director of IPP (since Schlüter had stepped down in 1973) for an independent assessment to be made of the design. Such a move was not even backed by the majority of the JET Scientific Committee nor whole-heartedly by the SB, but it was felt necessary to comply with the request and Bickerton's offer in May to ask the Reactor Group of the AEA at Risley to undertake it was accepted as the best solution. In that month the design features were collated into a 650-page report, R5 so fulfilling the requirement set down in the Design Phase Agreement that before June 1975 a detailed design of the whole facility was to be produced.

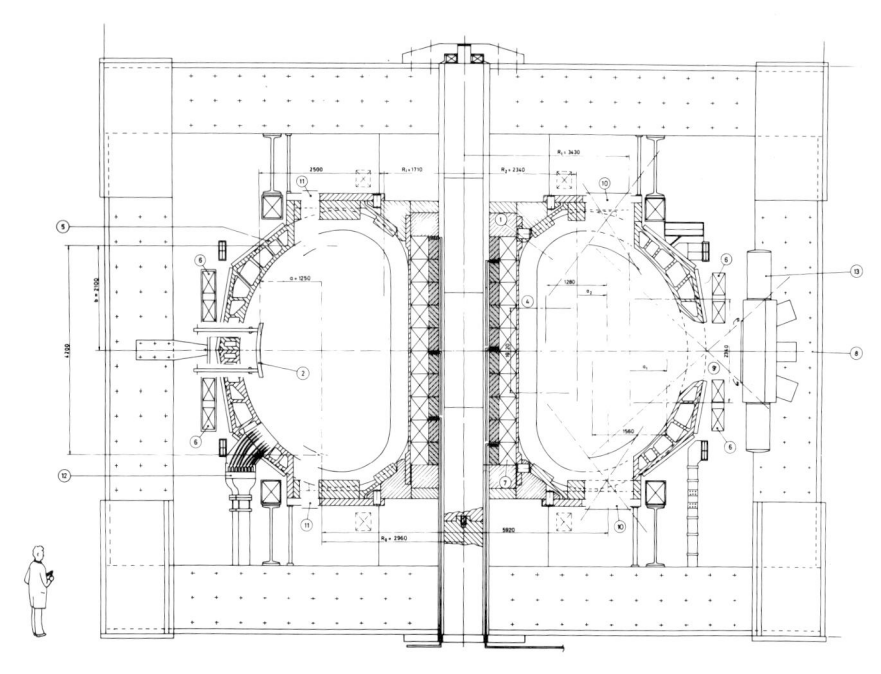

1 Vacuum Chamber	6 Poloidal Field Coils (External)	11 Vertical View Port (Small)
2 Limiter	7 Poloidal Field Coils (Central)	12 Cooling Pipes
3 Bellows Shield	8 External Magnetic Circuit	13 Cryo Pumps
4 Toroidal Field Coils	9 Radial View Port	
5 Structure	10 Vertical View Port (Large)	

A cross-section of the JET device as presented in the report R-5.

Independent Assessment

The 'Advance Summary of Conclusions' reached by Risley's Assessment Team and circulated in October 1975 was devastating, even though the oral presentation given by the leader of the team, Edward Hardwick, to the SB on 13/14 November somewhat softened the blows. As if this were not sufficient, results then being announced at the 5th IAEA Conference on Plasma Physics and Controlled Nuclear Fusion held in Tokyo put into question the fundamental basis of the whole design. It had been disturbing enough the previous year when the US Atomic Energy Commission had authorised the construction of Princeton's Two-component Test Reactor (which relied on high energy injection for a major fraction of the fusion reactions). Renamed the Tokamak Fusion Test Reactor (TFTR), the machine was to be of circular cross-section, modest aspect ratio, rather high field and a maximum plasma current of 2½ MA. Now it seemed the high field, high aspect ratio machine Alcator at MIT, was setting new records in the density × confinement time product and recently had been yielding temperatures with ohmic heating alone that were around

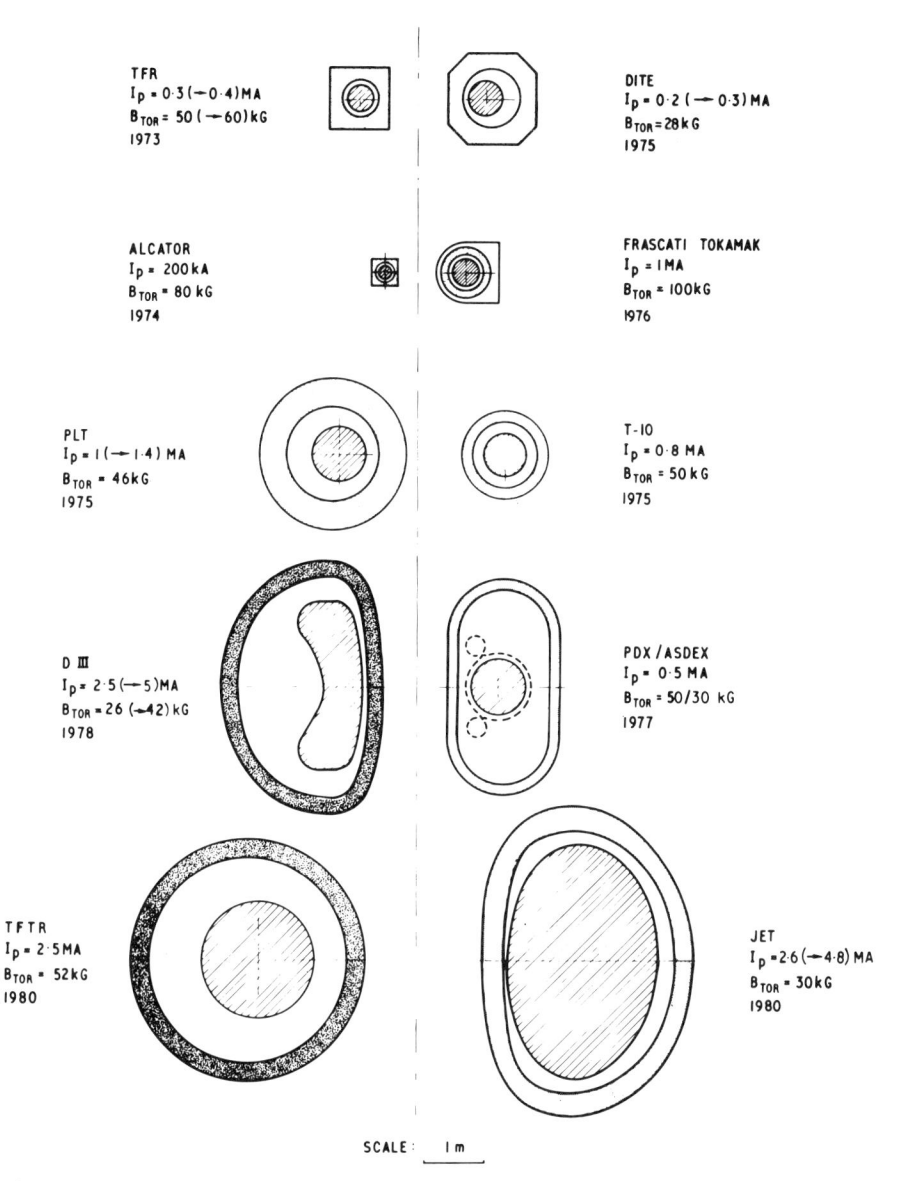

A comparison of the sizes and ratings of different tokamaks in the world.

10M°C. Alcator itself had had a chequered career, the vacuum vessel having collapsed in 1973 following an electrical short. Rebuilt in a simplified form, but with discharge cleaning, it went back into operation in January 1974 and produced plasmas of unparalleled purity. It reproduced the results obtained on Pulsator where IPP had shown the feasibility of building up plasma density during a pulse using gas injection, and discovered that confinement time increased as well and seemed even to be proportional to density. Moreover Alcator's results, and these seemed to be confirmed by new experiments in TFR, suggested that confinement time did not increase with current as expected and current was perhaps not the primary figure of merit. This was a complete reversal of the assumptions which had led to the big cross-section, low density, low field philosophy of JET.

It had been comforting to hear that at the international conference in Dubna held in July 1975, the overall maturity of the JET design in comparison with the equivalent American and Soviet designs had been recognised, which suggested that mechanically at least it was sound. It did not allay, however, all the misgivings over the European parameters, exemplified by Palumbo's remark made after the meeting at the end of June when the R5 report had been accepted that he still wished they were building a TFTR. Now the engineering was being called into question: Risley was stating that there were major deficiencies in the design—even if the basic concept was feasible—and after they had been put right, it would be necessary to allow six years for construction, not four, and add a further 23 MUC to the cost. Prevot's comment on returning home after the SB meeting was that if these criticisms were valid, then they would need to start again. It had not helped either to hear Rebut in the same breath talking of a peak power requirement of over 700 MW (whereas the first draft of R5 had cited 500 MW as the peak consumption) and treating the protests as if they were trivial.

To their credit, however, the SB rallied, took comfort from the preliminary nature of the report and instructed Rebut to assure the Team of its continuing confidence in their efforts. They were adjured to treat the report as a positive contribution to the development of the design and prepare detailed responses as a top priority. On this latter point the SB had to insist, otherwise Rebut would have limited the Team's reaction to his own terse dismissal. In January the full report was distributed to the Partners with a letter from Toschi concluding that even if the SB were to accept the findings, it 'would nevertheless strongly recommend continuing with the project as already presented' in R5.

In part the conflict of views between Risley and the team was inevitable if only because JET was conceived in the context of a French inspired scientific development which might appear from the outside to make a lot of assumptions, whereas Risley approached the assessment from a staid British engineering viewpoint applying reactor standards everywhere and seeing detailed difficulties at every turn. This was most noticeable in the reaction to the design studies made by industry on the vacuum vessel—one of the few to go seriously awry as exceptionally only one study contract had been let and the first designs produced (by Babcock and Wilcox) had been to fission reactor standards where long-term integrity according to the codes was paramount. Their proposal had been roughly rejected by Rebut as grossly too expensive and the procurement and construction time of eight years grossly too long. Such an approach however was seen by Risley to be correct rather than that proposed by the Team and subsequently developed by MORFAX. This closely followed Rebut's own conceptions and calculations, would be much quicker and cheaper to build, and in Rebut's estimation responded more closely to the requirements of a fusion experiment. While Rebut himself will still contend that the assessment exercise was a waste of time and that most of the points had already been taken care of — or would have been soon after — others see it as an essential interjection which at the very least obliged the Design Team to take stock, put the drawings in order and make sure the assumptions about being able to design out of certain difficulties were indeed justified. Whatever the stimulus, in practice important improvements were made in areas that had been identified as weak in the Risley report and some of Risley's estimates, for example on the time needed to develop, test and finally to assemble the vacuum vessel were to prove not so far out in practice.

The position however at the end of 1975 when the Design Phase Agreement was formally due to terminate was that the full report had still to be circulated and the two sides had still to get together to produce a consensus evaluation. In regard to cost, time scale and some essential mechanical features, therefore, the machine had to be regarded as subject to serious doubts. The advance report was supposed to be confidential but in practice the news leaked rapidly, bringing an angry reaction from Wienecke when the Directors met in December as he feared that the whole enterprise had been put in danger. At this meeting it was Trocheris who was the leading pessimist, suggesting that JET should in any case not be considered as a tritium experiment at all, while it was Palumbo who insisted that there was now no going back from the essential parameters agreed at Erice in September 1974. He had personally opposed them then, but now the Partners had to stick to their guns and take collective responsibility for the decisions.

Negotiations on Structure

In the meantime good progress had been made on obtaining acceptance by the laboratories of the administrative structure appropriate to building the machine. The report of Horowitz, Meusel and Willson provided a basis and the Design Phase Agreement provided a precedent. This, it will be recalled, specified the Committee of Directors acting as a broadly-based general overseer and the real control being exercised by a Supervisory Board of seven members (one each from France, Germany, Italy and the UK, two nominated by the Commission, and one to represent the interests of Belgium, Denmark and the Netherlands). In addition, a Committee for Scientific and Technical Matters would provide independent advice on scientific questions and a Committee for Administrative Matters on financial and administrative affairs.

The first draft agreement for the construction phase that Palumbo prepared for Schuster was written on the assumption that JET would be built by the Commission and its Associates in one of the laboratories of the Associates. It foresaw as supreme authority the Assembly of Partners which would decide the budget whereas 'management' as such would be the responsibility of a Board that could be the Committee of Directors. The word 'management' seems to have been given a rather special and not always consistent connotation in the evolution of the JET agreements which has persisted even to the present statutes. In this context it meant final right of decision on big contracts (although not signature which would be the responsibility of the Commission) and authority over the Project Head. In discussing the terms with the Associates, Pease and Braams had indicated that they would support the idea of the host Associate looking after contract administration; Wienecke, on the instructions of his government, was pressing for a top council at government level (not necessarily exclusively for JET); Italy, Belgium and Denmark were insisting on 100% Commission responsibility, with full representation of all interested states in the decision making bodies. (In 1972 Italy had rebelled against the form of Euratom representation on the Dragon Board of Management, which was, in consequence, enlarged to permit delegates from the Euratom member states to participate directly).

However, it rapidly became clear that the Committee of Directors was too big to act as a board of directors and the next proposal was a linear structure with an 'Assembly', 'Steering Committee' or 'Supervisory Board' of about seven people (advised by a Scientific Committee, an Administrative Sub-Committee and a Local Committee). Under this was the Head of Project and close behind the Project Board which Meusel would have merged into a single Directorate on a pattern similar to that adopted in the Max-Planck Institutes. According to this scheme, power of decision for contracts would be given to the Head of Project up to 20,000 ECU, Project Board up to 60,000 ECU and the Supervisory Board over 60,000 ECU. To avoid each contract passing through the full internal administration of the Communities, the General Directorate responsible had given written confirmation that powers of signature would be delegated to Palumbo for all contracts except the very biggest. Appointing a prime contractor was still not completely excluded although the SB had decided against; Braams remained in favour and Prevot feared it might be the only solution to the staffing problem.

A major point of contention was the position of the smallest partners who resented the private discussions going on in Brussels with the big three, typified by the submission of the scheme to the Atomic Questions Group before there had been full consultation on the Committee of Directors. After a number of exchanges, a compromise was reached to extend the Board to nine members but limit the voting powers to seven, although this would need to be modified if the explicit request of Sweden and the tentative discussions with Switzerland should lead to their becoming Partners. Similarly an adjustment might be needed to the budget contributions which were settling to 80% by the Commission, 10% by the host laboratory and 10% by the Partners.

By April 1975, agreement had been reached on both the principles and the wording. The Assembly and Steering Committee had merged into a Council, there was to be a management committee with weighted voting powers (two for the big, one for the small) and the preamble was to emphasise that JET was a joint venture, in which the laboratories of the Associations were closely involved, run under a regime that made it distinct from the host Association. The legal form was left open, awaiting the site decision with the choice lying mainly between an Association Contract (on the lines of the Design Phase Agreement) and a legal entity in the host country without its own staff.

For the Project this agreement had immediate significance as the Group for Atomic Questions had refused to consider further funding (particularly for prototype testing) until it was concluded. The Group was satisfied with the proposals and following its recommendation, in May 1975 the Council of Ministers, whilst stipulating that no longer term decision was implied, approved an amendment to the 1971-75 fusion programme, to implement the Design Phase Agreement by releasing the 2 MUC for long-term delivery items, a further 1.25 MUC for additional staff mobility and 0.6 MUC for the 20% additional contribution by the Commission to work for JET going on in the Associations.

Site Studies

The third element in the equation — the site — was in a much less satisfactory state. Recommendations were to have been made to the Council of Ministers in September 1974 for their approval in November. Schuster's idea was for a top-level independent committee to decide between the merits of each of the sites proposed whereas written into the Design Phase Agreement was the statement that the Partners would set up a committee representing all the Associations. As a result the members who were nominated in April 1974 were senior (but not the top) men from inside the various Associations and representatives of the Project Board, with Palumbo as Chairman. Vice-Chairman was Gunter Grieger of IPP, elected to the the post he believes because he was the only one who came to the first meeting with ideas on how they should proceed. His view, and he was supported by the other members, was that the Committee should analyse essentially technical aspects only and this was broadly the basis on which they worked.

Invitations to propose sites were sent to the Partners together with a questionnaire at the end of May 1974 and at the end of September, seven candidates had been received. Cadarache and Grenoble in France (the latter being withdrawn a little later), Garching and Jülich in Germany, Culham in Britain, Mol in Belgium and Ispra in Italy. Of these, Garching, Jülich and Culham were certainly fusion centres, Cadarache was a fission reactor experimental centre of the CEA where it was proposed fusion activities should be concentrated in the future, Mol a nuclear research centre, in part Euratom funded, and Ispra the main laboratory of the Communities' Joint Research Centre. Neither Mol nor Ispra had any significant fusion programme at the time, whereas from the very beginning it had been understood that JET would be built at one of the Association laboratories.

There are by now different versions of how the mandate passed to the Site Committee came to be interpreted and at what point the Associations through the Committee of Directors or the Groupe de Liaison or even the SB lost their grip on the situation. Probably at the time there was a breakdown of communication, not least in Brussels itself. Unquestionably a major factor was the Commission's decision to arrogate to itself the choice of site, once the project had been approved by the Council. The lack of realism in this approach is almost too astonishing to accept and it was certainly not Schuster's original assumption. No doubt it was partly conditioned by the Commission's general policy to try and regain some of its dwindling authority and it would not be a unique case of the Commission looking into its statutes and taking a stand on a questionable legal interpretation regardless of the political practicalities of the situation. Partly it was an underestimation of the importance that JET would assume, but whatever the background, the effect was to insulate each level from the next. Guido Brunner, the Commissioner responsible for fusion, was remote from Schuster, Palumbo felt himself cut off, the Directors understood they were not to interfere in high level politics which was out of their competence, while the Site Committee pressed on blithely with the majority following Grieger's lead that they were there to consider the largely physical parameters set down in consultation with the SB and the Project team.

The Site Committee grouped round D. Palumbo (in the chair). From left to right: D. Willson[+], M. Neve de Mevergnies (EB), M. Longo, (CNEN-CNR), N. A. Gadegaard (DEA), K. G. Stoecklin (KFA), E. Bertolini[*], J. P. Poffé[*], C. Gourdon (CEA), A. van Ingen (FOM), G. Grieger (IPP), C. Lafleur (EUR), R. S. Pease[+]
[*]*JET representatives and not members of the Site Committee*
[+]*Culham Laboratory representatives and not members of the Site Committee.*

Members of the Committee visited the sites during December 1974—an entertaining if not necessarily enlightening operation. Starting from Culham with which most members were familiar, the party moved on to France, where a helicopter was laid on from Marseille to compensate the lateness of the incoming aeroplane. At Cadarache, senior officials gave a particularly comprehensive introduction to the region—not fully appreciated perhaps as it was all in French when the Committee's working language was English. Travel on that night to Ispra proved impossible because of a strike and Garching advanced all its preparations by a day to cope with a revised schedule—showing an adaptability in which they took great pride. One of the fond memories of the trip was of Charles Lafleur, the devoted hard-working Secretary of the Committee (as well as of the SB, Committee of Directors and the Groupe de Liaison), 'reviewing' the firemen at Cadarache, lined up with their engines under the Sun. Similar to Palumbo in many ways, nervous, a heavy smoker, and utterly devoted to the cause of European collaboration in fusion, Lafleur's death in January 1976 was seen by the whole fusion community as a tragic loss.

Despite their different national or local allegiances and the influence these had on judgements, the members of the Committee were able to deliver a unaminous report in February 1975. Their analysis, summarised in the Table, indicated that all sites were acceptable, but the one which came out on top with two 'excellents', one 'very good' and one 'good' was Ispra, followed by Cadarache, Jülich, Garching, with Culham and Mol in bottom places.

	Technical Requirements	Radiation Safety	Support Facilities	Social Aspects
Cadarache	Good	Excellent	Good	Very Good
Culham	Very Good	Fair	Good	Fair
Garching	Good	Fair	Good	Very Good
Ispra	Excellent	Very Good	Good	Excellent
Jülich	Very Good	Very Good	Very Good	Fair
Mol	Fair	Very Good	Fair	Good

Unrated as unnecessary: Scientific and Technical Environment, Possibilities for Expansion!

The placing was, of course, somewhat superficial, since if unequal weightings were given to the different categories the result could be made to look quite different, and a little closer reading showed that comparisons were being made in some cases, between what could be with what was. Perversely the one crucial technical characteristic about which the Project could do nothing but which would directly influence project costs and the possibilities of expansion was, out of delicacy for national feelings, played down. This was the strength of the local power network. In the site specification, a capability of providing 300 MW over 30 seconds every 10 minutes had been cited and all could meet this, but only at Culham and Ispra could the full machine load be accommodated as a pulse while still leaving capacity to spare. At any of the other sites, the cost of meeting JET's steadily growing demands for power would have been significantly higher. In the event had the Site Committee contended itself with a simple statement that all the sites were technically feasible, a lot of trouble might have been saved.

As it was, the Commission was presented with an assessment that seemed to give a clear lead to its own site — a site for which they had been seeking a raison d'être for years. Without going into detail, Ispra had become the respository for lost projects from the ill-fated reactor system ORGEL onwards. Despite the efforts of Italy to divert programmes of real work to the centre, Ispra had remained a scientific and technical pariah. Every research programme evaluation presented by the Commission was liable to be threatened by the pretensions of Italy on behalf of Ispra on the one hand and the intransigent opposition of its partners on the other. Brunner was to face this problem all over again for the 76-80 pluriannual programme — a programme in which fusion it must be recognised was given relatively low priority. What choice had he? At the very least the report could be used as a lever to get some fusion work to Ispra for which the Italians had been barracking for a long time.

When Schuster learned, some time in the late Spring of 1974, that Brunner had every intention of making Ispra the Commission's choice, he was aghast. By this time he had become personally committed to JET going ahead and he knew there was no possibility of Ispra being accepted by the three big countries. By the Autumn, senior staff in the Associations also knew what the proposal would be and those in the large countries knew it would never be accepted. Each level, however, was powerless to intervene until the proposal was made formally. Whereas at the working level, there was little reserve — Ispra is in a magnificent location high in the Italian lakes, with a lot of space, plenty of power and water, the wine good and holiday resorts, both winter and summer, in easy reach — within the top echelons of government and science, the hostility to Ispra, in Germany, Britain and above all in France, bordered on the paranoiac. It had two roots: First, the site history — difficult access (because of either fog or strikes), low output, low morale, high wages; the non-professional staff employed by Euratom were paid up to three times the rate of those receiving local wages and labour relations were, in consequence, apalling with militant unions appearing to run the management. Second, the bureaucracy of the Joint Research Centre with all important decisions referred to Brussels and regular imbroglio with Community politicking. In no circumstances would the heads of government departments responsible for big science in France, Germany and the UK or the leaders of the CEA and the UKAEA (the parent bodies of fusion research in France and the UK) allow a high technology, large-scale novel project such as JET go to Ispra. Schuster knew this and in the last analysis so did Brunner, but the site report gave him little room for manoeuvre. Being German he could not appear to be favouring Garching as against Culham or vice versa, and he was more concerned with trying to negotiate his overall programme for the next five years. This no doubt explains why, despite repeated appeals from the Project, the SB and the Committee of Directors of ever mounting anxiety and frustration, no statement had been made by the time 1975 ran out.

As the Design Phase Agreement formally drew to its end therefore, JET had an internal structure agreed for the construction phase, but no legal form and no head of project, no site and a design plus costing that was subject to severe criticism. Nor was the rest of the fusion programme faring too well. In March 1975 the Groupe de Liaison had approved a draft pluriannual fusion programme of the Community for the years 1976-80. This would be the first subject to a new regime whereby every three years a five year programme would be established in order to avoid the hiatus that had always been likely to arise at the end of any period. The proposal had been passed for comment to the Scientific and Technical Committee of Euratom and also to the CREST committee. Whilst the former could approve the scientific content CREST gave warning that fusion was in competition with other sciences and some pruning would be necessary. CREST (the Committee for Scientific and Technical Research) had been a brain-child of Schuster who hoped by creating a high level committee of this nature to avoid programmes once agreed being mutilated by COREPER which had no feel for R & D. It was a forlorn hope as in Italy especially the foreign office is not prepared to be by-passed by other bodies, and so programmes still had to run the gauntlet of COREPER and its subsidiary, the Group for Atomic Questions.

Regardless of the unfavourable economic climate, the fusion programme that had been put forward anticipated a major increase in expenditure in the coming years — not just for JET but also for the national parts of the fusion programme. Whereas in 1975 the overall expenditure in Europe was estimated to be between 72 and 80 MUA* per year, depending on how much non-supported work was included, the projection for the five years 1976-1980 at March 1975 prices was a total of 615 MUA (123 MUA p.a. average), of which 135 MUA was for the four years of construction and one of operation of JET, and 480 MUA for the rest of fusion. For the Commission this implied 89 MUA for general support (at 25%), 58 MUA for preferential support (at 45%), 6 MUA for undefined work at Ispra (at 100%), 4 MUA for mobility contracts and administration plus the 108 MUA for JET (at 80%).

The Group for Atomic Questions predictably found this excessive, and it threw into relief the inherent limitations of the Groupe de Liaison which consisted of fusion people, who were highly qualified technically, but who in the majority were remote from government and science policy making, and whose executive powers were strictly limited. Particularly with JET now coming to the fore, the Group for Atomic Questions proposed that a European Fusion Council be formed of more

---

*The UA replaced the UC in most (but not all) official documents for a time.

senior scientific administrators who could look at fusion including JET as a whole. This had been strongly advocated by Horowitz who had grown progressively more and more impatient with the Groupe de Liaison which he considered at times irresponsible.

The Council finally came into being in April 1976 with the title of Consultative Committee on Fusion. Its membership included Jean Teillac, Chief Commissioner of the CEA, Wolf-Jürgen Schmidt-Küster from the German Federal Ministry of Research and Technology, Carlo Salvetti, Vice-President of CNEN, and Walter Marshall, Deputy-Chairman of the AEA and Chief Scientist to the British Minister for Energy, as well as Schuster from the Commission — a powerful group indeed, on a par with CREST, yet entirely concerned with the European fusion programme.

JET AGREEMENT

Despite all the uncertainties in the JET proposals, the Member States of the Communities signified that they were ready to approve construction at least in principle and also the means of financing this. The Communities would contribute 80% of the cost, the host country 10%, the remaining 10% coming from the national Partners according to a scale determined by the repayments they were receiving from the Communities in support of their home programmes. Cuts made in the European fusion programme insisted on by France, Germany and the UK were not too draconian and JET escaped unscathed. The Ispra allocation was removed to another budget, general support was reduced by 10% and only preferential support really suffered—cut by one third. The implications for the overall programme was a reduction from 615 to about 550 MUA.

The programme definition for 1976-80 accepted by the Council of Ministers included the full sum for JET which meant that if the outstanding problems could be solved, the way was then clear for construction to begin. However, the programme that was adopted by the Council on 24 February 1976 contained a commitment for 124 MUA only and excluded JET. It also set a limit on expenditure in 1976 of 20.8 MUC pending a final decision on the Project. Nevertheless, the essential decision whether the Community countries wanted to go ahead or not had implicitly been taken. No significance should be attached to the programme for the period being agreed after that period had already begun. This was not at all unusual, indeed the reverse. The Commission would prepare the programme well in advance and it would then do the rounds of the lower committees so that by the end of the year preceding the period concerned the final outcome was more or less known. Also the practice of up-dating every three years a rolling programme of five years meant that in future a programme could roll on at least as far as running and committed costs were concerned. The system, however, has one major deficiency. As no explicit provision is made for inflation, a planned profile of expenditure cannot be imposed and the Commission may spend all its capital allocation in the first three years in the hope that new money will be available in the fourth year. The temptation to mortgage the future, especially in times of high inflation, is difficult to resist.

In the popular image, the only major problem holding up the start of the construction phase was the choice of site, but France in particular had serious reservations over both the technical state of the design and the legal framework — determined at all costs to keep the project out of the hands of the Commission.

Technical Design

Devoid of political complications the quickest to resolve was the engineering. Over the first quarter of 1976 the Team prepared its response to the 250-page report that Risley had produced at the turn of the year. In part, it was a rebuttal, in part an explanation of the design work that had gone on since the compilation of R5, and in part an account of changes that had been made in the light of the report (even if not stated as such). The vacuum vessel had evolved considerably in its design and detailed stress analysis with the help of Saclay had confirmed its stability. Although Risley remained to be convinced upon the degree of vacuum that could be achieved, experience at CERN was compelling and in JET, where hydrogen gas would be injected, it was the partial pressures that mattered. Consequent upon a decision to increase the strength of the poloidal field, the mechanical structure was being redesigned in steel as against the light alloy indicated in R5, and the problems raised by Risley had already been accounted for. The vessel supports had also been modified to cater for wider manufacturing tolerances. Further testing and analysis of the coil designs had revealed no snags that could not be dealt with even though manufacturing procedures would need close attention. The consultant's criticisms of the power supplies had centred on JET's proposal to place so much reliance upon rotating machinery and to commit themselves to machines that were outside normal industrial experience. However it was accepted that JET had to put up a scheme that could work at any of the sites, including those connected to networks of limited capacity and it was conceded that the peculiar demands of the poloidal field might justify the flexibility rotating machinery gave. The arguments over the consultant's recommendation to use resistance switching as a method of profiling the power demand was really a minor point and was not pressed.

At a joint meeting held at Culham on 29/30 April a final exchange of views took place and a joint statement issued. Risley accepted that most of the points they had raised had been taken care of, that they had been unduly pessimistic in regard to the long term testing needed for the main coils and that Rebut's time scale and costing were not unreasonable. The only disagreement outstanding was the extent to which maintenance could be performed remotely, as Risley believed that the exchange of a complete octant might well not be practicable if only because of excessive time and cost. In any case, the Project argued, a high level of radioactivity requiring completely remote handling would in itself indicate that JET had already largely achieved its purpose and the time would have to come to move on to greater things.

The physics background was a little less reassuring. On 26/27 February 1976 a special workshop had been held at Culham to discuss the implications for JET of the results emerging from the high density tokamaks, notably MIT's Alcator, IPP's Pulsator and Fontenay's TFR which had been exploring high density discharges. These indicated that confinement time went up with density to a maximum that increased with the toroidal field. Electron temperature certainly rose with current, but rather slowly and on both counts any hopes of getting near the fusion regime through current heating alone was forlorn. The conclusions presented by two of the Team, Claudio Pellegrini and John Sheffield were that the key to approaching fusion conditions lay in providing massive amounts of additional heating. Without this, JET and Princeton's TFTR could be expected to have comparable performances, but less heating power would be needed for TFTR than JET because of its smaller size and higher toroidal field. JET's preoccupation with big dimensions and high plasma currents should give place to a concentration on developing systems capable of injecting, say 50 MW of additional power into the plasma. The amount forseen in even the extended programme was quite inadequate.

Such criticism from within, of the direction in which JET was headed, indicated that the Team was not quite as monolithic in its ideas as might have been presumed. Such opinions were also taken seriously. However the Project Board concluded that JET and TFTR were equally valid optimisations of the known data against cost and one could even imagine a third solution that looked like an enlarged Frascati tokamak. Where JET scored was in having far lower stresses in the toroidal field coils and the large aperture which gave flexibility in operating with different shaped plasmas. There was to be no turning back, but the importance of additional heating was acknowledged, and it was to allow operation at higher pressure, that the poloidal field was upgraded.

Site Negotiations

The repeated failure of the Council of Ministers to decide on the site for JET received a great deal of publicity and was cited by the Communities' detractors as a typical example of the nationalist posturing of politicians to which Europe has so often degenerated.

The Commission's choice, in pursuance of the thesis that site selection was their prerogative, was not promulgated until 21 January 1976 although it was to have been announced at a meeting of the Council of Research Ministers in December 1975. This was however, a bad-tempered affair and Brunner preferred to leave it to the February meeting. He justified the Commission's decision by asserting that previous experience of fusion was not regarded as essential, and placing JET at Ispra would cause least disturbance. Moreover, an international school for children of the staff already existed.

If the Commission really wanted a decision, the tactics could not have been worse. There is nothing wrong physically with Ispra — indeed a poll at Garching made amongst the staff in January 1976 gave preference to Ispra if Garching was not selected. The Design Team was also favourable and in a poll taken at Culham in May 1976, Ispra came out top of the list of preferences followed by Cadarache, and then Culham and Garching almost equal. The Site Committee was impressed by the enthusiasm of the staff they met there and the ready co-operation of the region, even to the point of simulating a pulse load of 300 MW by suddenly cutting off the supply from a hydrostation, just to confirm the estimated effect. Italy was, of course, enthusiastic and only Belgium of the smaller countries was against.

So much the worse, because as already stated, France, Germany and Britain were determined that JET should not go there. The UK view was strongly influenced by the AEA's anxiety to assure a future for Culham, whereas France, already doubting that the construction of JET was a matter of urgency, saw in Ispra and the Joint Research Centre the epitome of what they most abhorred in the Brussels bureaucracy. Brunner's assurances that JET would be run as a separate entity from the JRC failed to move them. The argument used publicly by the three countries was that Ispra was not a fusion site and so could not provide the back-up support the new Project would need. On the necessity for this, people are still divided. JET has not run into serious scientific or technical trouble from which Culham has been able to extract it, but it has had a problem of staff shortages and has depended a great deal on Culham for the supply of middle grade technical staff. Ispra might have been able to do as much and the Commission might have accepted a structure that gave JET real independence. Certainly salary costs would have been very much higher, but then only in Britain could the Project benefit so much from low rates of pay in the host station. At a site in Germany or France, the professional wage would have been little different even if less skilled workers would have earned rather less.

Of the other contenders the Commission argued: the move of the French fusion effort to Cadarache had still to take place, Garching had a stellarator and tokamak programme that would take their full efforts and Culham was busy with its work on a high beta machine. Mol had received little support outside Belgium and the claims of Jülich had not been pressed, although in terms of available power supplies, broad experience and ability to handle radioactive equipment it could have claimed to be the better German site. Jülich however as a collection of institutes with a collegiate style management was overall not keen to receive the new Project and institutes not directly involved in fusion feared they would find their own resources diverted into the fusion pocket.

Brunner continued through the Spring of 1976 to act as if he believed that in the long run, JET would be allowed to go to Ispra, and in March, Willson was drawing up an agreement based on this assumption. Willson had moved to Brussels in May 1975 following the appointment in February of Peter Oates as his successor, and attracted by the invitation of Palumbo to help establish the Project at whatever site was chosen, had taken early retirement from the AEA. Discussions in the newly created Consultative Committee on Fusion (CCF), whilst confirming the hostility of the three biggest countries to Ispra, significantly produced no alternative solution. It was at this time that another candidate site appeared on the horizon. At CERN, construction of the Super Proton Synchrotron (SPS) was nearing completion and senior staff were contemplating a future which might hold no more big machines. CERN had developed the most advanced pulsed power supply system in Europe, and had a vast experience in magnet technology and in the administration of big international construction projects. Against this background the idea was born in discussions between Adams, his deputy Hans-Otto Wüster, and John Fox, the originator of the pulsed static supply technique, that JET should be built at CERN, where it could take advantage of an existing international infrastructure. Bernard Gregory, former Director-General of CERN and then Director-General of the French Conseil National pour la Recherche Scientifique, was approached and expressed enthusiasm. The idea was also put to high officials in the German government, during a mission to Germany (to soothe ruffled feelings about CERN problems) in the Spring. But by this time they learned it was too late. The argument that it had to be a fusion laboratory to exclude Ispra had been put too strongly for another non-fusion laboratory to be considered. CERN's name continued to crop up from time to time, but essentially the site was never a serious contender. It is doubtful whether sufficient power, additional to CERN's own needs could have been provided and too much European research funding was already going into that corner of Europe; only France was sympathetic.

By taking no decision over JET in February, the Council of Research Ministers were at least able to resolve the conflict over the budget of the rest of the fusion programme, circumventing the Italian veto on coming to a decision while the site for JET had still not been determined. At the same time, JET became firmly entangled in the arguments over the JRC and Ispra's future. As if to complicate procedures, the JRC works on a 4-year programme and the next period was due to start in January 1977. The year 1976 was thus the principal period of battle and Brunner clung to the hope that he might solve the two problems at the same time.

An attempt to separate the issues was made by Germany which proposed that the Council of Foreign Ministers at its meeting in July 1976—the June meeting of the Research Ministers had been cancelled because of elections in Italy—agree that the JET Project should, in principle be accepted into the framework of the fusion programme. Moreover, the Research Ministers should, at their next meeting, approve the Project, subject to the CCF and the Commission having come to an agreement on outstanding problems. On 19 July such a proposal was agreed—with the caveat on the part of Italy that the site would be Ispra.

Far from all being then easy to resolve, the CCF at the beginning of October was forced to admit that it had not been able to arrive at any consensus and it could give no useful advice on the site. Despite the proclamations that if only the decision could be left to the scientists all would be well, neither at the level of the Supervisory Board, the Committee of Directors, the Groupe de Liaison nor now the CCF could a solution be offered. The Site Committee had evaded the issue and even at the level of the Design Team there was dissension as people either covertly or overtly campaigned for their own preference. Rebut, for example, had no wish to see it go to France, where the head could not be French, but then he was opposed to Culham and delivered an impassioned denunciation of the site to the Supervisory Board on one occasion.

By the turn of the year 1976/77 progress had been made in defining the programme for the JRC, and in view of the implacable resistance of the three biggest countries, Italy had given reason to believe that it would withdraw its veto on any other site provided a genuine programme of fusion technology was assigned to Ispra. In November the Council had obtained agreement that no one would use a veto if the others were unanimous and it approved by a large majority the motion to consider in future only sites with a long experience in fusion. This left the choice to either Culham or Garching and it was generally believed that the majority favoured Culham, including by then the Commission. Cadarache nominally remained on the books as being a future fusion site, but the preoccupations of the CEA were primarily with fission reactors and the Government was not willing to sink political capital into what they saw as a marginal issue, and not keen to make the necessary investment in the infrastructure in the region.

There were hesitations over Culham however. Marshall at CCF meetings had been treading on toes by intimating that only if the AEA were to run the Project was it likely to succeed. He had returned to advocating as construction policy, contracting out to the host laboratory the responsibility for building, and reducing as far as possible the international content of the management. This was strongly opposed by the smaller countries in particular, for whom the main preoccupation became the maintenance of the international status of the Project and the real participation of the whole European fusion community in it. Marshall was also guilty of the heresy of publicly doubting that CERN was much of a success.

This was the background to the meeting of the Research Ministers on 30 March and 1 April 1977. Gerald Kaufman of the UK was in the chair and the UK was represented by Anthony Wedgwood-Benn. Mario Pedini, the Chairman of the ill-fated meeting in December 1975, gave the discussions an encouraging start by announcing the first discharge in the Frascati Tokamak and Italy's willingness to go with the majority on the JET site. Brunner also indicated the Commission's readiness to accept the majority view and Hans Matthöfer, for Germany, made a similar gesture. France's position was that it had no objection to either Culham or Garching but it might reserve its assent until the structure of the Project had been approved.

All seemed set then for a decision on the site in principle at least. Questioning on the conditions that would govern the provision of both the land and services and the status of personnel in the country was led by the Netherlands, seeking assurance on the continuing international character of the enterprise. Matthöfer's replies were unreserved — whatever the Community considered desirable in terms of staffing and management could be accommodated at Garching, the terms of lease and services would be generous and a new international school was already planned. Much less satisfactory were the wordy and evasive answers from Benn who seemed incapable of making any commitments, with the result that opinion swung progressively in favour of Garching. Kaufman who had spent two months in bi-lateral discussions with the other countries and had been assured by Marshall that Culham had sufficient support saw the comfortable majority he had counted on being whittled away, and when Benn insisted on an 8:1 majority in favour of any site other than Culham, Matthöfer made the obvious counter and insisted on the same for Garching. Protracted discussions in restricted session failed to resolve the stalemate, although it did bring agreement on the JRC programme by all but Benn who declared a need to refer it back to London, thereby pigeon-holing the programme for a further number of months. Alienation of the British was thus further intensified and if a vote had been taken then it is perfectly possible that Garching would have got an 8:1 majority. Kaufman read the position and despite protests closed the meeting without putting it to the vote.

Matthöfer was furious and his resolve to urge the Garching case was correspondingly strengthened. In the early stages of discussions on JET, Germany had not been particularly excited about the Project. Wienecke, when he returned to Garching in 1973 to become chief scientist, had initially been preoccupied by initiating a reorganisation of the Institute's programme so as to concentrate on a few main lines of research rather than on the broad diffuse areas that had been current until then. The main instrument of research in tokamaks was the relatively modest machine Pulsator and plans were made for the much bigger successor — ASDEX. Wienecke, however, came to the conclusion that it would be good for the site if JET were to come to Garching and with his characteristic thrust and energy began to promote the idea. Within Garching, there were those who feared that such a massive project would destroy the university spirit of intellectual freedom and democracy that existed there but he soon had the majority of his people behind him. Next to convince was the Bavarian government, which was responsive to arguments on the industrial value to the region (particularly in competition with Nord-Rhein Westphalia) and in turn put pressure on the local electricity network to agree to accepting the pulsed high power load the Project would demand. Tritium handling would present problems but Wienecke was able to persuade the state safety authorities to agree that permission would be given to operate with a tritium filling if this should ever prove desirable. With this behind him he proceeded to bombard the Federal Government and finally was able to establish a personal relationship with Matthöfer and so obtain his support.

Whereas for Britain gaining JET was vital to bolster a faltering fusion effort, for Germany the issue became one of principle. Nor was the cause of compromise helped by the understanding in diplomatic channels that the French Government doubted the utility of JET and was in no hurry to take any decision at all. At the meeting of the European Council of the Heads of State on 29/30 June 1977, JET was not even on the agenda and the only useful outcome for fusion was that Britain shortly afterwards withdrew its block on the JRC programme. This at least helped to de-fuse some of the resentment amongst the uncommitted States.

Fortunately, Chancellor Schmidt and Prime Minister Callaghan were due to meet bilaterally and although the initial dates were postponed because of a terrorist kidnapping, they finally came together on 18 October in the euphoric atmosphere that followed the successful storming of a hijacked Lufthansa jet at Mogadishu in which Britain had been of help. It was the day after the Foreign Affairs Ministers had finally agreed on the structure following two years of almost continuous discussion (*q.v.*). A popular story of the Schmidt-Callaghan meeting was that out of gratitude Schmidt withdrew Garching's candidature, but this does not quite fit the facts. Certainly when the Research Ministers met on 25 October 1977 with only JET on the agenda, the crucial step of insisting on an 8:1 decision by either side had been withdrawn, but Garching was still in the lists. It should also be noted that over the previous months the Chairman of the Council of Science Ministers, Henri Simonet of Belgium, until recently Vice-President of the Commission, supported by his Prime Minister and Council President, Léo Tindemans, had been carrying out an extensive lobbying of his European colleagues. Taking a personal interest in achieving a positive outcome he concluded that positions were no longer as rigid as they had been, and a compromise was possible. It involved a partial agreement in regard to the next machine even though some Ministers reacted with apprehension to the very mention of a successor. Finally, Ministers were able to agree in three stages that:

1. A decision on JET 1 did not imply a commitment to build JET 2 — presumed to be a follow-up reactor experiment;
2. The site for JET 2 could be in any country other than the country in which JET 1 was located and could be at a JRC;
3. A majority vote on JET 1 would be accepted.

In that vote, influenced by the fact that no JRC was on British territory, and that 40% of European fusion work supported by the Community was currently in Germany, five voted for Culham, two for Garching and two showed no preference. Culham then it was.

During all the period of uncertainty, Rebut and his team through the complicity of the Commission, the Associates and industry, had pressed on with improvements to the basic design, tests of critical components, and filling in detail to prepare as far as possible for construction. As time wore on, however, and people—including group leaders and senior personnel—left the Project, the strain on the overseas staff became intolerable. For the British—largely from Culham—they were subjected to the same uncertainties in regard to careers, but they did not have the same problem of detachment from the home base, nor the difficulties that a family experiences working in a foreign country for an undefined period. Rebut had to contend with his own problems and also try and support his colleagues whilst at the same time being consumed with impatience and exasperation at the apparent incompetence of the politicians. At least he knew he could rely on the support of the Commission—Brunner himself on one occasion came to Culham to give encouragement—and the Partners (with the possible exception of France) not to mention the European Parliament and the Communities' Scientific and Technical Committee. All were anxious that the Project should be approved and all wanted the Team to remain together for as long as possible. The various extensions of the Design Phase Agreement—to 30 June 1976, 31 December 1976, 30 June 1977, 30 September 1977, and the prolongation of personal contracts went through virtually without opposition. The decision of the Management Committee at the end of September 1977 to start implementing the closing down phase was the result of an initiative taken by the Team on the grounds that the Project had become unmanageable. This can be interpreted as in part a means of exercising pressure, in part an expression of frustration and in part recognition that the steady decline in numbers was reducing the effort available to below the threshold level.

The measure, nevertheless, was real enough. Members of the Team had set about arranging their return—the jobs they would go to and the places they would live in. So close to the point of disbandment was the site decision, the Huguets for example had all their possessions in boxes at the time and were waiting for the removers. Most of the others from the Continent would have gone too at the end of October.

Rebut was vigorous in his campaigning from early 1976 onwards, writing to ministers and the Commission to urge them to come to a rapid decision in order that continuity would not be broken and bewailing the lost lead over the American and Japanese 'competitors'. He was amply supported in this by his senior staff who because they were in JET were no longer bound by the protocols that would have muzzled them had they been in their previous positions. Led by Bertolini and Gibson (who received a rap over the knuckles from Palumbo for their pains) a coloured brochure 'The JET Project, Status July 1977', was rushed out and sent to the Ministers in Brussels while telexes were sent to the foreign offices of all the Community countries. Nor were the wives silent, Mme Huguet sending a particularly appealing letter to the different foreign ministers at the end of September 1977, pointing out the problems of the non-British families. The staff were also active in pressing for the choice of a site which had sufficient personal amenities that it would attract the best people. Four aspects were identified: education and the need for a European school; housing, particularly for rent; jobs for wives; international status that would protect their pensions and minimize taxation. Whether any of the lobbying made any difference at all is difficult to evaluate. Certainly at the level of the Management Committee there was no lack of sympathy with their views and in the different countries the heads of laboratories were equally vigorous in their efforts to get decisions taken and have agreed a structure that was equitable and compassionate.

The site decision had a tremendous impact on fusion throughout the world. All had been working in the shadow of an impending collapse and the debilitating effect had been felt everywhere. Bertolini heard the news by telephone from Toschi whilst at an international fusion conference in Knoxville. At the end of the current lecture (by Paul Reardon on the TFTR) he asked for the floor and announced that JET would be built at Culham. Spontaneously the 1000 participants rose to their feet and burst into emotional applause.

P.-H. Rebut (squatting) with members of his team photographed round a model of JET soon after the site decision. From left to right: P. Dokopoulos and D. Eckhartt, K. Selin, J.-P. Poffé (behind), P. Noll, B. Green, E. Salpietro, J. Last, G. Celentano and A. Gibson.

Structure Arguments

In effect decisions over the project structure and the site were taken at the same time implying that both problems took equally long to resolve. In Brussels, the tendency was to see the arguments over structure as a dalliance pending the decision on the site, but this was grossly to underestimate the strength of feeling in France, particularly over the need to prevent the Commission having any control whatsoever over the running of the Project, whilst the small countries were nervous of the big countries excluding them from full participation unless Brussels played a central role. It is equally legitimate therefore to contend that arguments over the structure removed the sense of urgency from the site question.

Throughout the discussions Palumbo advocated the formula of Association Contract that had been the backbone of the development of a coherent fusion programme in Europe. As already stated he had received written assurance that the relevant Directorate was ready to delegate the right to approve contracts to an Executive Committee and Palumbo was ready to delegate much of his signing power to Commission staff on site. No novel legal instruments would be required. The Groupe de Liaison had already agreed on a management chain so that a smooth transition from design to construction could have been immediately instituted. In the final form, the chain had comprised a Council, a dedicated management committee, followed by Head of Project and the Board. The work of the Supervisory Board set up to oversee the design phase was finished once the questions on the design had been settled and this was signalled by acceptance at its May 1976 meeting of the report on the exchange of views at which JET and Risley had come to terms. The Consultative Committee on Fusion breathed a sigh of relief and called for the early establishment of the Council and the JET Management Committee (MC). For the time being, it was decided the CCF could act as Council, and it was only necessary to form the MC. This met for the first time on 7 July with a membership that looked very like that of the old SB which had been wound up at its 26th meeting in June, enlarged to include delegates from Belgium, Denmark, Ireland and Sweden. Pease was elected Chairman. Whereas the SB had had a narrowly prescribed mandate, the terms of reference of the MC were not so restricted and it became an important forum for exchanging views on detailed structural points, although its first task was clearly to promote acceptance of the Project and prepare the way for the construction phase.

Foremost amongst the pressing needs was the appointment of the Project Leader and at that first meeting a sub-committee was formed to determine the appointment mechanism. Subsequently Schuster and Pease together drafted an advertisement announcing the post and this was sent for publication to *Europhysics News*, thereby provoking a major storm in Brussels where it was interpreted, correctly, as a form of public pressurising. The Journal had however gone to press and no withdrawal was possible. The Commission had to accept the fact and support it with good grace. The response was impressive even though irrelevant;

certainly no new appointment could be made in advance of the site decision and the MC had in any case 'urged Rebut to act further as if he were sure to be the leader of the construction and exploitation phase also'. The phrasing was ill-advised, but it should be recalled that the MC saw itself as the body responsible for JET decisions within the broad policy laid down in the CCF, with Rebut and the Project Board acting as executors. So far the scale of JET had not really penetrated and at this level the machine was thought of as another DITE, ASDEX, or TFR—a little bigger of course, but not different in kind.

Within the CCF, under the steady pressure from Teillac who saw it as his bounden duty to take the Project out of the hands of the Commission, the need for it to have its own legal identity became accepted, even if reluctantly by the smaller countries. France had made clear during 1975 even that it would never accept such an enterprise simply as an inclusion in the fusion programme. Schuster understood and accepted this and in mid 1976, instituted serious studies on just what the implications of different forms of legal entity in different countries would be. Of central concern was the status of the staff. At that time the Head of Personnel within the Commission believed (wrongly as it transpired) that if JET were a legally separate body it would also have to be the employer of the staff engaged on it, whereas secondment from the Associations had been the accepted principle since the earliest discussions. When working out the Ispra site proposal, the concept had been respected by providing for employment by the Commission at Commission rates when the staff worked at JET and imposing an obligation on the Associations to re-employ when the contract with the Commission ended—the return ticket as it came to be known. On the other hand neither France nor Germany favoured employment by the Commission, (although there had never been any difficulties with the Euratom staffing at the JRC laboratory in Karlsruhe) preferring the ILL system where staff are employed by the legal entity at rates comparable to local rates. Such a scheme would nevertheless not be acceptable at Culham because of the low salaries paid there; allowances would need to be greater than the salaries and recruitment from abroad would be almost impossible under those conditions.

Comparison with ILL can be misleading. ILL was set up to provide experimental facilities around a specially designed reactor for a wide variety of short-term users coming from different research areas. The reactor is not the object of research, but a unique tool and the users may be working on quite different topics. The Organisation began its existence with two partners only and was grafted onto the structure of the Centre d'Etudes, Grenoble, with scarcely any disturbance. In contrast to JET, there are no home-based equivalents of the Associations whereas JET was essentially conceived as an extension of the Associations. Consequently when the question of the secondment from the Associations came to be discussed by the MC, the concept of guaranteeing a return ticket was readily adopted, and subsequently it was never seriously questioned.

At the beginning of August 1976, following the rather optimistic meeting of the

Foreign Ministers, Brunner issued instructions that the target for reaching conclusions on the structure should be the meeting of Research Ministers in October. The Commission was to remain neutral in the argument over staff conditions and as so much opposition had been voiced to the principle of establishing JET under an Association Contract, a serious look must be taken at making the project into a Joint Undertaking to preserve the Community character. Schuster sounded out the reactions of France, Germany and Britain to the idea and received cautious encouragement. He then introduced the notion to the MC at its second meeting on 1 September as being a goal to aim at whatever initial form the entity took. It was a middle road between the big countries calling for an independent limited company (or its equivalent) and the small countries plus Italy still promoting the Association Contract. Only then did the idea of creating a Joint Undertaking *ab initio* emerge, yet the status as such had been mooted as a possibility in Palumbo's earliest review. The explanation lies in the interpretation that had traditionally been placed on Articles 45-51 of the Euratom Treaty. Characterising an enterprise as a Joint Undertaking was regarded as an instrument for giving a Community character to a national industrial activity of indefinite duration, which was of value to the Community as a whole. In return for information, certain privileges including relief from taxation were accorded. Yet the wording of the Articles in no way makes this restriction and there were ample precedents in adjacent fields; the majority of the big projects of the European Nuclear Energy Agency had been set up as Joint Undertakings. Slowly the members of the MC came to see the possibilities already inscribed in the Treaty and finally agreed to setting up a sub-committee under Luc Ornstein of the Netherlands to study the question further. Guided by Karl-Hans Melchinger, who even before becoming a member of Palumbo's staff in 1975 had been made responsible for drafting the agreements involving fusion, the sub-committee weighed the pros and cons and finally came down firmly in favour of creating a Joint Undertaking. The MC at its meeting on 22-23 September, still with some misgivings, endorsed their conclusions. The principal hesitation concerned the need to seek the Council's approbation of its statutes with the risk that the terms would become another object of long drawn-out argument and extortion.

When the CCF and then the Council of Research Ministers in October 1976 agreed that JET should be a Joint Undertaking it seemed that a great step forward had been made — and so it had. The terms of the JU had still to be worked out, however, and not all issues were devoid of controversy. The question of voting power in the Council, for example, became a source of contention — Horowitz especially feeling strongly that votes should reflect the scale of contributions (although not for the Commission of course) and he received support from Britain and Germany. Subsequently this has been shown to be a quite unnecessary provision for in common with other international scientific organisations experience has indicated that weighted voting brings nothing of value, but can provoke a lot of resentment. More difficult were such questions as the disposal of assets and liabilities once the Project terminated — a source of great dissension towards the end of the Dragon project — because what starts as an asset turns into a long-term radioactive liability. More serious was the inter-relation between the Partners and the Commission on the one hand and the Partners and the Head of Project on the other.

At the level of the CCF and above, Rebut's qualities as scientist and engineer were appreciated but he was not felt to be the right man to be overall director with responsibility for all the administration, contact with the Partners and the Commission, preparing budgets and so on. His task was to complete the design, supervise the actual construction, and make JET work. Names had been exchanged over the years of possible candidates for the post of Director — not all of whom it might be added, proved suitable on a closer examination. In Germany, the ministry concerned turned its gaze on CERN where one man came to the fore as having the right qualities. This was Hans-Otto Wüster, a convinced European, an accomplished scientist, a most able administrator with an in-depth knowledge of big international science and big project control, of unimpeachable integrity, and strong. He was not a fusion man, coming instead from high energy physics — but science policy makers saw this as not necessarily a disadvantage. If the site was to be in England, with Rebut being French, a German director was strongly indicated. Moreover, tentative probings indicated he might be interested.

Despite the disastrous outcome of the meeting of the Council of Research Ministers on 30 March/1 April 1977, a high level approach from the German Government was made to Wüster in April, as a result of which he went to Bonn to talk over the possibilities, returning with the draft statutes to study. Two weeks later he turned the idea down, as he put it later 'I was not prepared to hold the steering wheel while others controlled the accelerator and the brakes'. When Gregory, by then Délégué Général for scientific and technical research, heard of the refusal he dispatched one of his principal aides to find out the cause; Kofoed-Hansen from Denmark also came to enquire. Wüster convinced them of the impossibility of running a complex innovative project when detailed decisions remained in the hands of a committee whose members had other loyalties and knew nothing of the in-house problems at first hand. Control there had to be, of course, but either Council defined the financial limits or the programme. It could not do both at the same time.

In the light of his experience as Director-General of CERN where he had been solely responsible to Council for the implementation of an approved programme and budget that he himself had prepared, Gregory was totally sympathetic to Wüster's view. As principal French delegate in CREST and close adviser to the Minister for Science and Technology he was also involved in parallel with Teillac (whose terrain was the CCF) in the fight to construct the statutes so as to give the

Project greater autonomy. His reaction was to make representation at the highest level in France where he received the undertaking that France would not permit the Project to go ahead unless the management structure corresponded more closely to that of CERN. Of all the countries involved, France at this time — middle of 1977 — was the least concerned about the possibility of JET being abandoned. JET was seen as an interesting scientific experiment — one among others — but probably ahead of its time. Fusion had been rushed along unnaturally fast and many thought the accent should still be on plasma physics research. Also JET was liable to divert effort from the change-over from Fontenay to Caderache which was becoming the CEA's main preoccupation at that time and the Project was neither likely to lead to a system of immediate commercial prospects nor provide a great deal of work for French industry. As a result France was ready to go to the limit to ensure a structure that it considered right and a steady pressure was exerted at the level of COREPER in pursuit of this goal. At the same time an effort was made to hold down staff numbers and budgets, and generally trim the intended investment.

The first battle to be fought, however, was for the absolute supremacy of the JET Council in JET affairs, and for control over the Project in between the agreement in principle and the final approval of statutes for a Joint Undertaking on a specific site. France insisted that this had to be the Council ad interim and not the Commission on a sort of Association Contract. By June 1977 they had won these points and the next issue was the role of the Executive Committee. In the draft agreement examined by COREPER late in September, one particular phrase was in dispute, notably that allowing the Council to give the Executive Committee 'the task of supervising' the management by the Head of Project if Council were unable to do this! Such a provision went right against the principles enunciated by Wüster and Gregory and early in October the provision was finally changed to — 'the task of following the financial management of the project'. Moreover in addition it was stipulated that the Head of Project 'is responsible for management' and 'answers to the Council'. COREPER approved the draft including the overall cost and staffing on October 13 as a point 'A' implying that it could go through the Council of Foreign Ministers without discussion. This it did, as we have seen, on 17 October 1977.

Head of Project

At the end of that same month the decision on the site was taken and Wüster was once again approached by the German government with an invitation to apply as Head of Project — under his rules. He accepted.

For many who had struggled so earnestly over so many years to have the Project approved it was a victory tinged with bitterness. 'Our common enterprise', although named a Joint Undertaking had become a separate unit, set off from the existing fusion community, distanced from Palumbo's family of Association Directors (who were seeing their own budgets suffering because of it), and liable to pass into the hands of a stranger. Above all for Rebut it was a cruel blow when he learned from Teillac as soon as the site was chosen that he was not to be the overall Project Director. It needed all of Teillac's patrician authority — a quality that was to be of great service to JET in the years to come — to persuade him that his real role was as Technical Director and it would be a waste of talent for him to become an administrator however senior.

Rebut had borne the brunt of the years of uncertainty, driving the Project on, refining the design and anticipating as far as possible the availability of construction funds. It was he who had kept the Team together, reduced in size though it was, and had provided the leadership that had given the design its essential coherence.

Financial Provisions

Although it was not until the end of 1977 that the major hurdles were cleared, the Project had not been left to languish without funds. Delays in making them available caused perturbations, but the effects were minimised by Schuster's and Palumbo's ingenuity and the relaxed approach of the Associations towards reimbursement, notably Culham. As a result, significant sums were at the disposal of the Team for studies and prototype work, and although these would have been largely wasted if the Project had not gone ahead, this never was made an issue. Explicit provision had been made in the Design Phase Agreement even in the Aim of the Project 'to prepare and place, if so decided, the order for certain long-term delivery items'.

Rebut began pressing for the release of the 2 MUC foreseen for this in the Spring of 1974 and the Supervisory Board approved the proposal at its meeting in Erice in September of that year at the same time that it accepted the broad specification set out in R2 and determined the basic budget figures. However, the Groupe de Liaison was not prepared to go so far, the chief opponent being Horowitz, on the grounds that the placing of initial contracts would prejudice too large a fraction of the subsequent construction contracts (a consequence very thoroughly discussed before the Design Phase Agreement was drawn up). Pease also was against as he wished to see the structure defined first, being still at the state of advocating the appointment by Euratom of a prime contractor for the work. Palumbo, however, aware of the necessity for having the allocation approved within his 1975 budget got agreement that the money should be voted even if its release had to wait agreement on the structure. By March 1975 the Groupe de Liaison had come round and was urging the Commission to obtain the release of the 2 MUC as quickly as possible and in May the Group for Atomic Questions concluded that the structure was sufficiently agreed for the money to be handed over.

For Article 6 work in the Associations it was not money that was the main problem but finding the people to detach on to JET oriented work. When it came

Hans-Otto Wüster

to the point, the Associations were reluctant to divert too much effort with the consequence that progress on, for example, additional heating and diagnostics was slower than might have been expected. Even with the delay in the programme, JET started up with only the so-called 'basic' diagnostics in place and the supposedly principal auxiliary heating plant—the Neutral Injection System developed at Fontenay and Culham fell seriously behind schedule. Delays in both diagnostics and auxiliary heating arose from an under estimation of the technical difficulties and the variety of lines pursued, but the problems would have been more accurately assessed if the scale of effort in the Associations had been more in keeping with the size of the Project and if the momentum in the Design Team could have been maintained.

Measures to establish an interim funding for JET were begun in the Spring of 1976 when it was appreciated that a decision was many months away, the JET agreement having been extended without any new funds being voted. In September the ceiling for Article 6 contracts was raised from 3 MUC to 4 MUC, the additional Euratom contribution going from 0.6 MUC to 0.8 MUC, and in the following month Council agreed to the allocation of a further 4 MUC for expenditures not funded by the programme (excluding JET) in part to cover Article 7 contracts. Prudent management by Rebut and the Management Committee through 1976 and 1977 and the run down of staff to a total of 34 combined to limit current payments and keep long-term commitments in check, although even in September 1977 when it was implementing the closing down phase, the Management Committee was still authorising the extension of certain industrial contracts. Nevertheless, so carefully was the balance maintained of keeping the Project viable whilst limiting the cost, that come October 1977, 1.9 MUC was still unspent. Such judicious husbanding of the funds over the entire period would not have been possible had it not been for the introduction (and the widespread acceptance of industry) of the notion of staged contracts.

In October 1975, the Commission had organised in Brussels an Information Day on JET with the express purpose of awakening industrial interest in the Project. The Commission was also keen to promote industrial cooperation in fusion, generally looking ahead to the time when power generation became an essentially industrial process. In view of the uncertainties over JET already clouding the horizon, such an ambition was much too premature, but the presentations given at the meeting and the wide distribution of R7—a general introduction to fusion technology and the JET design—ensured that industry was aware of the scale and scope of the Project.

On Rebut's insistence, despite hesitations in the Supervisory Board, industry was consulted on major items usually well before the manufacturing phase. Design studies were used to explore ideas or alternative approaches and were not tied to production at all (although many members of the Supervisory Board and of the Groupe de Liaison remained unconvinced on this point). For many in the fusion

community, having been used to small machines, partly built in their own laboratories, or by local companies with a useful background in a particular technology and often on a cost-plus basis, dealing with large-scale manufacturers chosen by international tendering was a novel experience. Equally, tokamak technology was a largely novel field for industry and the design studies were valuable introductions to both sides. On completion of a study contract, the JET Team would assume the responsibility for defining the design so that by the time it came to manufacture it was much more fully developed and analysed than was usual for research equipment. Contracts for manufacturing were divided basically into three stages:
1. Final design, production of drawings and any testing.
2. Tooling; prototype manufacture; purchase of material.
3. Actual production.

A contract covered all three stages, but only if and when the money became available could one go from one stage to the next. This was the risk that companies took and it was for them to decide whether that risk was acceptable. Through this system many of the big items on the critical path were well advanced by the time the Project finally came into being, notably: the delivery of the copper for the toroidal and poloidal coils, manufacture of the first toroidal field coil, first vacuum vessel bellows assembly, and the manufacturing design of the rigid sectors. Altogether 'orders' at firm prices had been placed for about 20% of the capital equipment before the Project was approved. Moreover the prices agreed compared favourably with the estimates included in the design report and this gave confidence in the reliability of the rest of the estimates. Fortunately under the Commission's rules, tenders can be called for before the financing has been agreed, whereas in many countries such a procedure would not be permitted.

In total the sums invested in JET amounted by October 1977 to some 13¼ MUC. Largest item in the account was the 3.77 MUC spent under Article 6 contracts of which the Associates paid 77.3% (a quarter of which was reimbursed by the Commission as general support) and the Commission 22.7%. Article 7 contracts to industry, paid entirely by the Commission, came to 3.6 MUC, and salaries, split 77.3%:22.7% between the Associates and the Commission, to 2.6 MUC, to which should be added the Commission's payment of 1.7 MUC in overseas allowances and subsistence, and 0.2 MUC for Euratom staff as such. The main item remaining is 1.4 MUC for services at Culham of which the Commission paid 75% and the UKAEA 25% approximately.

ENLARGEMENT OF THE FUSION COMMUNITY

By the time the Council of Ministers had approved the basic structure and determined the site one further Partner had joined the nine members of the Community and a second was waiting in the wings. Soon after the energy crisis of 1973/74 rocked the economies of western Europe, the government of Sweden, anxious to demonstrate that it was not just standing helplessly by, applied to the Communities to join the fusion programme. The request received only a lukewarm reception because at that time Sweden had no co-ordinated fusion centre and such research as there was, was carried out in the Universities, mainly those of Stockholm and Göteborg. Moreover, Stockholm had made clear that it would not be jumping on the tokamak band-wagon but would continue to pursue a broad line. This dispersion of effort made the conclusion of an Association Contract in the traditional sense rather difficult. Contracts were, in the main with specific laboratories or their parent body, and although in the case of Belgium the State was the Associate, only two specific university departments were involved, one of which was State-run, and communication with the leaders of the scientific effort was straightforward.

Whilst the JET Supervisory Board indicated its readiness to accept Sweden already in April 1974, the other committees concerned with the European fusion programme required more persuasion. Early in 1975 a delegation from the Committee of Directors went to Sweden to study their activities at first hand and not all were convinced that they could be usefully assimilated into the Communities' fusion programme. The Groupe de Liaison was equally reluctant and demanded more information on what Sweden had to offer. One answer to that was very scarce manpower and this was probably the deciding factor that led to the signing in May 1976 of an agreement for cooperation in the field of controlled thermonuclear fusion and plasma physics. Sweden's participation in discussions preceded the formal signature by many months, a delegate having joined the Committee of Directors from September 1975. The Swedish Associate, the Swedish National Board for Energy Source Development, became a Partner in the Design Phase Agreement on 1 July 1976 and when the Management Committee was set up later that month a Swedish delegate was included. Sweden's position was thus recognised and it was accepted in the discussions over the JET structure that the Swedish National Board would be included as one of the Partners, its general standing identified by its voting strength, being made equal to that of Belgium, Denmark and the Netherlands.

Sweden has paid generously for the privilege. Its contribution to the Communities, calculated on the basis of its national product, has not been matched by the reimbursements made to its own programme (because of its modest size) as evident in its special contribution to the 10% of the JET budgets paid by the Partners which varied between 0.2% at the beginning to 0.14% in 1983. Few contracts went to Sweden from JET for development or manufacture, owing to the lack of interest or competitivity in the industry. The share up to the end of 1983 was a modest 0.32% whilst the share of the JET budgets taking into account its contribution to Euratom approached 3%.

Switzerland is in the reverse situation essentially because it has a centralised and much respected home programme and its industry has a lot of experience relevant to fusion technology. As far back as 1970, Palumbo approached the Centre de Recherche pour la Physique des Plasma (CRPP) attached to the Ecole Polytechnique Fédérale de Lausanne (EPFL) with a view to its becoming an associate of the Communities' fusion programme, but found little enthusiasm for the idea and negotiations were allowed to lapse. Fusion research in Switzerland had been initiated in the wake of the 1958 Geneva Conference with the setting up within the Fonds National of a laboratory for plasma physics research, located in Lausanne. Initial difficulties in finding a Director led to a slow start and for some time the total strength failed to rise above the threshold level. The quarters were in any case inadequate and, in 1964, it was decided to erect a special building as the Centre for research and this was finally occupied in 1969. Further problems arose as the Fonds National supports individuals and the Technical Director, Erich Weibel, found himself the virtual owner of the establishment. This brought him into conflict with the tax authorities and placed him in an uncomfortable position in regard to insurance for example. The situation was clearly unsatisfactory and after long and difficult negotiations, agreement was reached that the Centre should be attached administratively to the Ecole Polytechnique of Lausanne whilst still retaining privileged links with the Fonds National. Progressively during the 1970s the Centre became more integrated into the EPFL, but it still remains an extra-departmental unit, receiving also a modest income from the Fonds National.

Concentrating on plasma physics in the belief that fusion should emerge from the science and not be an empiric technology, the CRPP quickly gained a reputation out of proportion to its numbers and Palumbo saw it as a desirable addition to the existing Associates. It was not until the middle seventies, however, that discussions were renewed after Switzerland had concluded an agreement with the Commission for a partial integration of interests. This established a precedent for collaboration and at the instigation of the Department of the Interior, negotiations were reopened with Brussels. Fortunately it fell to Francis Troyon to provide the technical expertise in the place of Weibel who was on a year's sabbatical in the USA. Weibel was far less keen on a Swiss involvement particularly in JET. A complication was the condition laid down by the Government that, along with other activities, participation in the Communities' fusion programme was to be dependent upon the passing in 1977/78 of a bill introducing value added tax into the country as this was to be the source of funding. The bill in the event was rejected by national referendum and Switzerland's participation in JET might have ended there had the cause not been taken up also by Maurice Cosandey, the President of EPFL. Under his impulsion, the Government was persuaded to make funding available from other sources and despite the continued opposition of Weibel (by then returned), an agreement for cooperation in the field of thermonuclear fusion and plasma physics was concluded between Euratom and the Confederation on 14 September 1978. The necessary steps for Switzerland to become a Partner in JET were completed in August 1979, an 'entry fee' being paid to back-date the membership to 1 January 1978. The Government insisted on the agreements being made with the Confederation to leave open the possibility of other institutes taking part in the fusion programme. While the effective Associate for JET is CRPP in Lausanne there is, nevertheless, a substantial activity in fusion technology in other institutes, notably EIR and SIN now merged into the Paul Sherrer Insitute.

Like Sweden, Switzerland's contribution to the Communities is calculated on the basis of national income, whilst reimbursements are based on the overall fusion effort and the recognition of projects qualifying for priority funding. In contrast to Sweden it is rare that the return falls far short of the investment, as can be seen from the specific contribution to JET which rose to 0.43% in 1983 while its contribution to the Communities' Fusion Programme was running at about 3½%. Contracts placed in Switzerland up to the end of 1983 grossed over 5.5% of the total. As Associates, Sweden and Switzerland are exactly on a par with the other Associates and their programmes are coordinated in the same way as those of IPP or Culham for example. Their participation in the Communities' fusion councils as well as in JET is in every way similar to that of a full member of the Communities.

INTERIM PHASE

In the document approved by the Council of Ministers in October 1977 setting out the essential features of the JET Joint Undertaking it was explicitly stated that immediately after the site had been chosen, an Interim JET Council (IJC) was to be established. Its tasks would be to continue the supervision of the Project and, helped by the Management Committee as required, to prepare for submission within a four-month period the draft Statutes of the Joint Undertaking, preferably unanimously agreed. During this period the contracting parties were asked to agree to the Project continuing as a further extension to the Design Phase Agreement with the IJC assuming responsibility for its execution.

The IJC met for the first time on 23 November 1977. It comprised two senior delegates from each of the nine countries, (with the exception of Luxembourg) plus Sweden. Schuster and Palumbo represented the Commission. Teillac was elected chairman and the smoothness with which the IJC accomplished its tasks owes much to his calm and measured approach to problems, his ability to distill the essentials and present reasoned solutions which took account of national viewpoints, but still preserved the flexibility and independence of the Project. His own non-partisan approach established the constructive mood that characterised the deliberations of the IJC (and its successor the JET Council proper). The JET Management Committee was asked to carry on with its previous work until the Joint Undertaking came into being and three working groups were set up attached to each of which was at least one member of the Project Board. Statutes came under the chairmanship of Georg von Klitzing of the German Ministry for Research and Technology; Finance under Arnold Allen, Deputy Chairman of the UKAEA; Support from the Host under Braams, and in all three groups was Melchinger from the Commission. At its next meeting in December a fourth group, Staff, was set up with Palumbo as Chairman.

The first item of business at that second meeting, however, was the appointment of the Head of Project, strictly provisionally only pending the adoption of the Statutes. To a number of the Council, Wüster was known by reputation only and so a formal interview had been arranged. Moreover in case he should not prove acceptable, the IJC retained the candidatures of Toschi who had guided the Supervisory

Board through the Design Phase, and also Rebut. However, no last minute objections were forthcoming and Wüster was duly appointed Director (at the request of the Belgians, not Director-General) with Rebut as Deputy-Director and simultaneously Technical Director. Toschi was made chairman of the JET Management Committee in place of Pease who had recently resigned. At that stage a Directorate of four people was envisaged with department directors filling technical, scientific and administrative roles.

Statutes

Although a great deal of detail had still to be incorporated, the essentials of the Statutes had already been formally approved by the Council of Ministers, including the composition of the JET Council (two members nominated by each participant) and their voting rights (five each for the Commission, France, FRG, Italy and the UK, two each for Belgium, Denmark, Netherlands, Sweden and eventually Switzerland, one each for Ireland and Luxembourg) and the services the host laboratory was required to provide. Moreover, discussions notably in the Management Committee, had exposed the likely points of disagreement. Principal amongst these was the status of the JET Executive Committee, and Italy in particular continued trying to give it some real powers in the management of JET against the general will of the biggest countries and the absolute opposition of Wüster. During the debates in the IJC, Wüster, in general, kept a low profile, but in the corridors he let it be known that on this point he was adamant.

The Italian position was mainly a reaction to the neat distribution of functions amongst the big three — President of the Council, French; Director of the Project, German; site, British. They wished the Committee of which Toschi was Chairman designate to have equivalent importance. So strongly did they feel on this point, Pietro Longo came to the IJC with a complete revised draft of the Statutes and was prepared to argue them clause by clause. When it came to the point, however, Italy accepted a compromise that gave the Committee at least formal status and they withdrew their opposition to the other provisions. In the Article relating to the Organs of the Joint Undertaking, the Executive Committee was explicitly mentioned as being created to assist the JET Council, whereas the organs as such were defined as being the Council and the Director. In the interests of preserving the single line of authority, the suggestion that it assumed a special responsibility for collaboration with the Associated Laboratories was also not retained and its role made one of 'promotion'. As to its composition, this it was decided should reflect that of the Council and the voting rights as well. Individual representation from the fusion laboratories was rejected as distorting the balance, although Germany would have liked to have separate representation from IPP and Jülich.

The Director's position was further strengthened by rejecting the Commission's attempt to assume the prerogative for the selection and appointment of staff — this was the Director's function and not to be usurped at any price, even though formally the non-British staff were to be Euratom employees. The Council nevertheless would require the Director to seek approval for the appointment of senior staff.

The Commission, as providing 80% of the funding also wished to have a special status in regard to financial matters, but this was firmly opposed by the other Partners and the Commission did not insist. Voting rights in the Council equal to one of the large countries only had been stipulated in the document agreed by the Ministers and so was not an issue, but the modesty of the Commission in this context should not pass unremarked. Palumbo was a great advocate of the principle of fitting policy to the possibilities and Schuster was all too aware of the suspicion of the big countries and the spectre of another JRC. More surprising is the acceptance of Brunner and his colleagues on the Commission of this relatively minor role in the control of JET.

An interesting point of discussion, but not dissension, was the status of the JET Council in regard to other authorities in fusion. Its role of supervision was not in question, but for whom was this role being assumed. Was the Council in turn 'responsible to the Members of the Joint Undertaking' — a phrase that was proposed at one time? The conclusion was 'No' and it was decided that the Council simply would report to the Members. Nor would the Council act on behalf of the Commission and whereas the official seat was to be at Culham it was explicitly stated that the Joint Undertaking was not a UK company even though English law would apply in cases not covered by the Statutes. JET was thus to be set up as a unique enterprise whose existence depended entirely on the Council of Ministers.

The Ministers had already approved a duration for the Project of four years construction, plus one year commissioning and first operation, followed by five to seven years exploitation. In the Statutes this was made a total period of twelve years from the founding of the Joint Undertaking with provision for a possible prolongation. All Members were required to remain in the Project for the first five years after which withdrawal was permitted following due notice, a provision which primarily affected Sweden and eventually Switzerland. The UKAEA was not permitted to withdraw and as the other Members, would pay directly only a fraction of 10% of the cost (while their governments paid a much bigger sum as their fraction of the Commission's 80% share), unilateral withdrawal would be rather pointless.

One of the longest articles to be put together concerned information and patent rights, not because this was seen to be a major issue, but rather because the terms had to be fully compatible with the Communities' existing rules on the subject. All information generated by the Project would become the property of Euratom, but the Members evidently would be entitled to use it for their own research. For onward transmission to third parties Council proposed that it should determine the conditions as the need arose.

Whilst the Statutes of the Joint Undertaking was the key document that had to be submitted to the Council of Ministers, it was not the only one. Associated with it there had to be documents on:

- the scale and time-table for financing,
- a definition of the special advantages that would be conferred on the Joint Undertaking starting off with those presaged in Annex III of the Euratom Treaty,
- exemption from value added tax that would normally be levied on goods and services in the UK.

In addition, the Council would wish to have available the Finance and Staff Regulations and the Host Agreement between the Joint Undertaking and the UKAEA. Four months was not a long time for all these to be drafted and agreed — unanimously if possible. It says much for the application and the good will of the members of the Working Groups that they could finish in time. It also demonstrated the wisdom of having a single responsible body in charge — the Interim JET Council — because when problems arose, guidance was sought and the work could continue to flow.

Full advantage was taken of the change agreed at the beginning of 1978 in the money unit that was to be employed by the Communities from that July onwards and which would be based on real exchange values. In the early days of the design, the unit was the Unité de Compte, the value of which in different countries did not follow the same currency ratios as those corresponding to either the official exchange rates or the open market rates. Comparing tenders and working out costs in official UC would thus have introduced a totally artificial element. The SB worked around this and, on Palumbo's recommendation, from March 1970 used the Belgian Franc as its real money, converting this to Community money at the rate of 50 BF to the UC (or UA according to the custom at any one time). The base line for the costing was thus 135 MUA at March 1975 prices to be interpreted as 6750 MBF. This was updated by the Management Committee in February 1977 to 150 MUA at January 1977 prices on the recommendation of the Finance Sub-committee it had set up the previous December under the chairmanship of von Gierke. The Sub-committee however did not go so far as to accept Rebut's arguments for updating by 50% the 3 MUA Clause 7 money already released on the dual grounds of inflation over the additional year and his need to divert funds from long-term delivery items, in part to compensate the shortfall of secondments from the Association. They did however take note of the fact that the new 4 MUA being sought for long-delivery items was only worth 80% of its nominal value because companies had to be paid in real money and the Belgium Franc then stood at 41.3 to the UA. The strict inflation calculation had given only 149 MUA and the 4 MUA would eventually be accommodated from the main Project budget, so 150 MUA was seen as a nice rounding off.

This was the figure included in the document approved by the Council of Ministers in October 1977 although the programme this would cover had become a little blurred. Four years was still considered as the period corresponding to the construction period proper, during which some 137 MUA would be spent; the remaining expenditure would be incurred in the fifth year (after the termination of the current 5-year plan that ran from 1976 to 1980), and would cover some operation costs. Moreover, emphasis was placed on the fact that this investment would cover basic performance only and that some of the costs of extending the machine would already fall in the 1976-80 programme.

When it came to translating these figures into the new European Unit of Account, this was done by returning to old Belgian Francs and then reconverting to EUA at the current rate of 40.6 BF/EUA which raised the overall figure to 184.6 MEUA. At the same time a more realistic profile of 128.0 MEUA to be spent in the first four years and 56.6 MEUA in the fifth year was introduced. Again it was emphasised that these figures did not include the cost of extended performance and the statement that 'completion of the construction of JET and its auxiliary facilities, its commissioning and its initial operation within its basic performance will be included in the new programme for 1979/83' laid the ground for establishing that construction plus commissioning would take five years. These changes went through without protest and the overall project cost when later updated to January 1978 prices came to 204 MEUA, which included (exceptionally) an inflation figure of 14% for staff costs as under the newly introduced rules of the Communities, taxation on salaries was to be included in the budget.

Annex III of the Euratom Treaty sets down a number of privileges which a Joint Undertaking might enjoy and the only questions raised in their application to JET concerned the relevance of some of the clauses. For example, as the UKAEA already owned the land that was to be put at the disposal of JET there was no need for special recognition of its public utility character or for its expropriation. Similarly, as the Joint Undertaking was not a company, there would be no taxes on its foundation. The special privileges that were relevant however were readily accorded and the Member States agreed to waive customs duties and direct charges on equipment, property tax and restrictions on the import or export of scientific equipment or materials needed for the Project. They also agreed to allow JET to transfer money between Member States, a facility of vital importance to the Project as many of the Member States had very strict currency control regulations in force at the time.

In addition to these privileges the UK gave an undertaking that purchase tax already levied on goods and vehicles would be refunded and that JET staff coming from Sweden and Switzerland, as well as their spouses and families, would be treated as if they were from countries in the Communities. Staff members from other countries and their families would also have their stay facilitated.

Financial Regulations

Wüster, although not a formal member of any of the Working Groups and still resident in Switzerland, exerted an important influence on discussions. As Allen, the Chairman of the Working Group on Financial Regulations, was to say on a later occasion, 'within a week of his appointment... this vibrant, clear-minded articulate figure... (was) advising us how the project should be run... I came to realise that all of us in the Group were being guided, and wisely, by a newcomer of great experience and massive good sense'. Wüster's main preoccupation was to put more flexibility into the Regulations as they had been drafted by the Management Committee and limit the role of the Executive Committee to advising the Council without in any way diminishing the accountability of the Project to the Members. His regular references to CERN caused irritation, but the Group came to realise that whatever their opinion of the role of high energy physics in modern science or the particular scientific achievements of that establishment, a great deal of experience in handling a European project had been accumulated there and, in particular, the financial arrangements had been proved over many years. They could not simply be taken over because the Financial Regulations for JET had to be compatible with those of the Community and they also had to take account of the presence of the Associations which were contributors in their own right. A further difference relates to the indefinite life of CERN whereas at that time and throughout the construction period, JET was regarded as essentially a single project that would cease in June 1990. Nevertheless, CERN's major construction projects have been handled under a separate budget and whereas for example, the CERN general budget is adjusted for inflation in advance by an amount calculated according to a set formula, the major construction projects like the ISR and the SPS used a method of cost updating that was post hoc and related to real price increases.

In the original draft of the Regulations prepared by the Commission, the Commission was given specific financial control but this was not acceptable to the other Members and the Commission agreed not to press for more powers than those accorded to one of the bigger Members. Auditing of the accounts by the Court of Auditors of the Communities is in the nature of a service to all the Members. The Commission had also proposed a three-tier structure with an authorizing officer, accounting officer and control officer, but the others preferred the two latter functions to be combined so that the accounting officer assumed also the internal control. This officer is responsible to the Director of the Project who assumes the function of authorizing officer with powers of delegation. The Director has also reporting to him an internal audit service as a continuous watch-dog on the legality and regularity of the way in which the budget is implemented. The Regulations are very firm on the necessity for raising a requisition for commitment and having this authorised before agreeing to any charge on the budget, a measure deemed absolutely essential for keeping expenditure in check and firmly under the Director's control.

Much discussion was spent on how best to ensure a wide industrial participation in the Project without overloading the JET administration or wasting companies' money in preparing tenders that were not accepted. In mind was the prospect not only of the JET construction but also of the longer term involvement of industry in the exploitation of thermonuclear fusion for power production. Never seriously in doubt was the principle that contracts would be awarded basically on price provided the technical specification and delivery date could be met (as against the notion of 'juste retour'). The Commission, however, produced also distinctions between 'automatic tendering' (open tender awarded to the cheapest offer) and 'discretionary tendering' (limited invitations and the 'most attractive' offer accepted). This was not liked, as being liable to give rise, on the one hand to a lack of flexibility that could be technically damaging and, on the other, to the exercise of unacceptable pressure from interested parties.

During the Design Phase, satisfactory procedures for handling contracts had been drawn up on the insistence of the Committee of Directors, wherein the Partners acted as advisers on potential suppliers in their area, the Supervisory Board and then the Management Committee as controllers of the propriety of awards. If in the beginning there was an occasional lack of rigour in maintaining the confidentiality of quoted prices, or looking for alternative suppliers this merely reflected the small number of people handling a large number of contracts, many of which were at the limit of current technology, and the anxiety of the Team to get on as fast as possible. The Team had also to get used to the discipline of international tendering as for many it was their first experience.

For the construction phase an even more rigid system was needed and this was explicitly set down in the Annex to the Financial Regulations. In essence, JET is required to go out to competitive tendering without discrimination among the countries of the Members of the Joint Undertaking for all technical and scientific goods valued at 30,000 ECU or over and 10,000 ECU for others. All companies tendering are given access to the results of any previous design studies and if, during discussions with a company before tenders are in, new information of relevance emerges, or the specification is changed in any way, all tendering companies are informed. Contract terms and conditions are to JET's rules which further reduces the risk of disparities. The opening of the tenders is a formal occasion chaired by the Director or his representative, assisted by the Accounting Officer or his representative. Normally contracts up to a value of 150,000 ECU can be awarded by the Director without further consultation, those above are referred to the JET Executive Committee as well as those above 30,000 ECU when (for good reasons) the full international tendering procedure has not been followed. Should the recommendation of the Director be to place a contract with a company other than the one offering the lowest price (based on the exchange rate obtaining on the opening day) the arguments for the choice have to be convincing.

The Executive Committee proposes companies able to supply goods or services corresponding to different areas of work that JET has defined and is also given the opportunity of commenting on the list of companies that will be approached for a particular tender.

A natural fear was that an excessive number of offers would be prepared, but in practice this has rarely materialised. Typically three to six offers are received which is probably close to the ideal. The habit of asking companies to confirm their interest before sending out an invitation to tender is believed to have been an important factor in limiting the number. The distribution between countries has also proved to be reasonable if the very large contracts for civil engineering work are ignored. These almost always go to local companies and so distort the global figures.

When preparing his budgets, it was agreed that the Director should present the overall cost estimates for the entire duration of the Project, a forecast of annual expenditure for the following five years and an estimate of income and expenditure for the following financial year. This process was to begin early in the current year to allow ample time for examination by the Executive Committee; Members' acceptance or rejection was placed firmly in the hands of the Council, as was the level of staffing. It could not be ignored that the major contributor, however, might find itself without funds come the beginning of a new year in which case the Council could hardly adopt the budget. Provision was therefore made for the amount of new commitments into which the Director could enter pending agreement, to be limited to the figure already earmarked in the Members' budgets.

The financial year of JET was made to conform with the calendar year and its official currency was stated to be the ECU which is converted into other currencies at the current Commission rate. This raised a nice problem of banking in view of the existing UK laws which prohibited holding foreign currencies in bank accounts. No difficulty was foreseen in having an account in ECU in the UK, but for foreign currencies provisions had to be made for holding currencies overseas or in an external account. In drafting the Regulations much argument revolved round the question of whether payments made by the Members should be simply in equal fractions of the annual amount due or in sums that corresponded to the cash flow. In view of the smallness of the contributions of most of the Members, this would seem to be of quite minor significance, but perhaps this was the proverbial equivalent of the colour of the bicycle shed door.

No a priori decision was taken on the method of indexing prices although a number of delegates would have liked this to be inscribed in the Regulations. Instead, Wüster's recommendation was followed that this should be decided by the JET Council on a proposal by the Director when it came to deciding the first budgets. Provision was already made in the Regulations covering the presentation of the budgets that a clear distinction be made between increases that arose from changes in design, exchange rates and inflation, and on balance, it was felt better to trust the Council to monitor the situation rather than attempt to prescribe in advance.

Host Agreement

Vital to the efficient launching of the Joint Undertaking and the subsequent construction of JET were the arrangements to be made with the host laboratory. In outline these had already been defined in the document approved by the Council of Ministers prior to the site decision, but a huge amount of detail specific to Culham had still to be worked out. It had been laid down that the host had to provide at its own expense a site for the Project which met the 'standard conditions'. These included land of an appropriate area on which JET would be built and possibly extended, main services brought to the site boundary, all necessary permits in order and the roads for adequate access. In addition temporary office accommodation, laboratories and workshops were to be made available rent free for the first two years whilst permanent accommodation totalling 6000 m^2 was being built at the Host's expense for subsequent renting to the Project. Buildings specific to the construction and operation of JET were the direct responsibility of the Project. Power of 500 MW capacity plus a security line was also to be made available at the site, as well as appropriate cooling facilities and computing links. All the services required by JET such as workshop capacity, library, restaurant, office supplies and so on were to be provided at cost.

One of the most difficult problems was to readjust the attitudes of the Design Team and Culham laboratory towards each other. From being a minority group of visitors, the JET staff had acquired proprietorial rights and theirs would be the dominant project on the Culham site in the future. The Joint Undertaking was a thing apart from the host laboratory, as the delegations had again reiterated at the beginning of discussions, and its independence was to be guarded at all costs. This led to a certain truculence on the part of some of the senior staff who gave the impression that all they had to do was to demand and Culham was obliged to provide. They also viewed the efforts of Pease to integrate the two laboratories as far as possible as a covert attempt at take-over. Pease was conscious of the responsibilities he would carry for such matters as safety and if he had a temptation to see the Project as his own, this could be understood. He had been the initiator in the earliest days and much of his efforts over the previous seven years had been devoted to promoting the Project and Culham as the host site. One minor incident that blew up into a cause célèbre concerned the desire of Pease to link the two sites by a foot bridge over the dividing road. This was strongly opposed by the old Project Board and they were really incensed when Pease responded by reserving space for a Culham building to house certain common services on the far side of the road, linked to other Culham buildings by a bridge. The old guard of JET saw all this manoeuvring as symptomatic of Culham's acquisitiveness and a threat to JET's independence in the future.

In the Working Group, however, the desire of all parties to find equitable solutions and to ensure that JET could start up under the best conditions possible was predominant. Under Braam's guidance and with Willson in Brussels and Oates in Culham preparing the drafts, both fully aware of Culham conditions and the personalities involved, progress on preparing the Support Agreement went forward smoothly. Then when it was a question of finding a compromise between two distinct viewpoints, Wüster and Pease thrashed these out between themselves. Wüster coming in from the outside had none of the conditioned reflexes of his colleagues, yet was very clear that he was going to run his own show. He was fully capable of sparring with Pease with great skill and determination all the time in high good humour. Pease recognised an adversary of quality who was above petty score-marking for its own sake. His evident reasonableness and ability to see the essentials of a problem ensured that attention would be concentrated on matters of substance. In only the second draft of the Host Agreement prepared before Christmas 1977, it was foreseen that there would be a common security service and that people would use the same restaurant facilities and so on, so that useless duplication would be avoided.

Different interpretations were placed on the significance of the term temporary accommodation, the UKAEA choosing to understand this as describing the nature of the building and the JET negotiators as defining the length of occupancy. The difference was not trivial, as apart from the JET Design Team having already occupied permanent buildings in the Culham complex for a much longer period than expected, the relative advance of the machine components as compared to the buildings where such units could be assembled and where the machine would be housed implied that JET would be wanting a lot of storage space with appropriate cranage and this had not been budgeted by anybody. The compromise was for Culham to agree that office space provided in permanent as well as temporary buildings was to be counted as temporary and would be rent-free so long as it was needed, whilst Wüster undertook to limit the storage space he would require and agreed to pay for space from the end of the first two years. Storage space became something of a passion with him and any requests for expenditure on storage areas were later to be refused with a vigour that bordered on violence.

Another topic requiring clarification was the provision of workshop facilities should the machine tools normally to be expected in a research workshop not be adequate for JET purposes. Here, as in a number of other cases, it was left to the two parties to sort it out as the need arose. Culham was to provide a $600\,m^2$ workshop within its own premises equipped to its own standards and in which JET would have absolute priority. Should other machines be needed cost sharing could be discussed.

From the beginning of discussions, a Liaison Committee between JET and Culham had been envisaged although its role was changed from that of advising the Council to being the forum where matters of mutual interest are discussed. One of its specific duties is to decide which of three methods of payment would be applied to the provision of a given chargeable service:
1. proportion of the overall cost as agreed between the parties;
2. a tariff based on plant time and current man-day costs;
3. actual cost when a service was directly metered.

The Liaison Committee has served a thoroughly useful purpose, whilst the personal contacts developed between the JET and Culham directorates was to become a key element in establishing the ease of relations that evolved over the ensuing years. Symptomatic of this ease was the linking of the JET buildings on the new site to Pease's once-contentious intrusion and the sharing of the Culham Public Relations Department housed in it. Having served the JET team throughout the Design Phase, John Maple, the head of Public Relations, was to his considerable chagrin, informed in December 1977 that JET would do its own PR in future. But it soon became evident that JET people had neither the time nor the experience and Wüster, with Pease's agreement, asked him to take it over again, a move that JET had no cause to regret.

Rent of the non-specific buildings was of course a source of argument but finally the Authority's figure of 11% on the initial capital investment was accepted on the understanding that this would include full maintenance and redecoration and that after 20 years only the maintenance costs should be paid. Given the level of inflation and the interest rates obtaining in the UK, the figure was not excessive.

The terms which came in for most revision during the drafting phases concerned the provision of computer facilities. Initially these had been based by the Design Team rather narrowly on immediate demands, no doubt on the assumption that JET would become self supporting. Those finally agreed, however, were much more orientated towards long term needs and the steady growth over the first few years. One can in hindsight question the wisdom of planning to rely for big computing on the facilities at Harwell and their special links with Culham, but this was again a case of concentrating attention within the Project on the most pressing areas. Computing requirements could be divided into the same three phases as the Project itself. During the Design Phase computing was mainly needed for analysis and optimisation and was wanted on site while the staff was still in temporary accommodation. For the Construction Phase, the urgent task was to define the system that would inter-connect the myriad controls and sensors on the device and its auxiliaries so that tender action and software development could go ahead in time for it to be used for testing and developing sub-systems as soon as components began to arrive on site. Methods of managing and analysing the experimental data were on a longer time scale as they would only be needed in the third phase when the machine became operational. At the time, it seemed sensible to take advantage for this of the big computer and data storage facilities at Harwell, particularly as it would be for a limited period only. Computer services, however, have a habit of becoming saturated rather earlier than expected and JET's own needs for experimental data treatment and storage would have justified the purchase of dedicated facilities, even though they did not figure in the cost estimates.

The staff were naturally concerned with local amenities, not least with housing. Culham is a small village some 4 km from Abingdon, 10 km from Didcot and 20 km from Wantage—towns of 20,000, 14,000 and 8,000 inhabitants respectively, already servicing the 8,000 strong research establishments on the Harwell site. The two nearest cities, Oxford and Reading, are 12 and 30 km away respectively. Short-term accommodation was particularly difficult and the Authority provisionally took up a lease on Nuneham Park, an attractive estate with sports facilities that contained nine family flats and 39 bedrooms. To the Authority's disappointment, the Working Group declined to recommend taking it over and the option was allowed to lapse. The reason was that the Partners did not want to be encumbered with domestic property and were concerned by the expense. Despite the rejection, the Authority instead made available a number of dwellings under its control and has done what it can to help newcomers in their first months.

Schooling

From a very early stage the availability of a school where the children of the JET staff could be taught in their own language was seen as quite indispensable, and a big debate had gone on in England over whether it would be better to set up a special school or to use schools in Oxford. The decision was for a special school and to seek to have it classed as a 'European School'.

The first European School was recognised in 1953 in Luxembourg, a school that had started as a private venture and then been taken over by the State. Through its efforts, the European baccalaureat was created and in 1959 became accepted as qualification for matriculation. Since then it has become the rule that in order to encourage mobility, every European laboratory within the Commission's purview should have a European School near. These are run by a common Board of Governors which is totally independent of the local education authorities and has full powers to determine the conditions under which the teachers work in the different schools and the curricula followed by the pupils. It is the responsibility of the host country to provide buildings and basic equipment whilst the Commission looks after salaries and operating expenses. The teachers which make up the European teaching force are seconded by their home country—a process that is complicated only in Denmark and the UK where they are not civil servants even though paid from public funds. In the mid-80s there were altogether nine schools ministering to 13,000 pupils. The scheme is warmly appreciated within the Communities and the European Parliament is working towards establishing a European School in every major city.

Once it was decided to seek the European School solution, the UK authorities, under the impetus of Oates, lost no time in pressing forward with the project, and in March 1978 an application was made to the Board of Governors for a School to open the following September. This was granted immediately on 17 March. The UK was fortunate in having premises only a kilometre from Culham that were suitable and were being vacated, and these were acquired by the Department of Education and Science. Culham College, as it was called, was a Church of England college of education that was due to close in September 1979. There would be no intake in 1978 and so some accommodation would immediately become available.

The Culham International School.

Wüster saw not just the existence but also the quality of the School as being of capital importance for the success of JET. Only if the fusion scientists overseas were convinced that the schooling would be of high class would the best be willing to come and work at JET. Characteristically when the School opened on 18 September 1978, one of the first pupils to be enrolled was his own son, Wolfgang, which gave him a double incentive to promote the School's well being. Happily the man chosen in June to be the Headmaster was Derek Hurd, previously a headmaster in Abingdon and so familiar with scientists and their families and already an enthusiast for the European concept. Together they made a formidable lobby and they persuaded the British Government, after the conversion of the buildings to be more suitable for pupils of different age groups rather than adults, and the transformation of the residential part into classrooms, to modernise the laboratories and bring them up to a very high standard. JET also provided from time to time redundant equipment that could be usefully employed. Only once did the two men come to a confrontation and that was early in 1985 when the starting hour of the laboratories was advanced from 8.30 to 8.15 and with the backing of virtually all the parents (including those at JET), Hurd refused to follow suit. Not only was it inappropriate for the children's sake but it was even an advantage from the point of road safety to space the two times. Wüster was furious, thundering that the School owed its existence to JET. For once, however, he was forced to give way, and on this, as on many other occasions, it was left to George O'Hara, the Associate Director for Administration, to smooth over the troubles. Quiet, thorough, but not easily bullied, O'Hara was an admirable foil to Wüster's noise and bustle.

Whilst it was true that JET was the original motivation for the School, it was not a JET school and by that time was not even referred to in those terms as it had been in the beginning. Having started with 51 children between the ages of 4 and 14, by 1985 the number had grown to 750 and the School covered the full pre-University range. Children of JET parents however, constituted only 40% of the total, even though they had right of entry without payment and the majority of the JET staff took advantage of this. Entry rules caused a lot of dissension. No problem for the non-British at JET who were Euratom employees, but the British staff posed difficulties. Neglecting contractors, all the British on the site were employees of the Authority, yet some, as we shall see, were counted as JET staff, others not, even though they might be working on JET. The former had the right to send their children to the European School, the others not. But if Culham staff had been given right of access, the Communities would have been funding staff of a national laboratory. And if Culham, why not nearby Harwell?

Nevertheless, opening the School to other children was essential if it was to rise over the threshold in a sufficiently large number of languages. If numbers in a given section fall to five or under, the ruling of the Board of Governors is that the timetable must be reduced and some integration arranged. Early on this caused a lot of problems, notably in Italian. An immediate source of eligible pupils was the European Centre for Medium Range Weather Forecasting located at Shinfield near Reading, not close but well within bussing distance, and the Governors agreed also to take (for a small fee) children brought up in European languages other than English from a catchment area that includes Banbury some 50 km to the north and Swindon, about 35 km to the south-west. Language is the criterion for acceptance, not nationality. When an application is received from a British national who has been educated in another language it receives sympathetic consideration. Only at the point of entry does the question of entitlement or non-entitlement arise. Once accepted, all children are on an equal footing.

Hurd saw the role of the Parents Association as having been very important in the School's development. Scientists in general, he believed, are especially concerned by their children's education and continually need reassurance. This anxiety, however, can be channelled and they have a real contribution to make in formulating policy, raising money for special purposes and organising extra-curricula activities. At Culham even the transport for the pupils is run by the Association.

At first there was some nervousness over the standards that could be reached, but when in 1982, 100% success was achieved in the first year that pupils entered for the European Baccalaureat and again in the second and the third year, the reputation of the school was assured. It is interesting to record that of the eight pupils who got their Baccalaureat in 1982, despite English having been taught as a second language, seven opted to go on to an English University (among them Wolfgang Wüster).

JET Staff Conditions

From the very beginning of the Project it has been accepted that the staff would be representative of the European fusion programme and would in large measure be seconded from the Associated Laboratories, returning to their base once their work for the Project was complete. JET was to be a fixed term enterprise comprising a construction phase, a tuning phase associated with bringing on the equipment needed to achieve full performance, followed by a relatively short operational phase in a radioactive mode. Staff commitments would therefore be of limited duration — even though the meaning of this was far from fully analysed. It was also rapidly agreed, once the idea of creating a Joint Undertaking had taken root, that whilst many of the staff would be employed by the Communities as temporary servants filling temporary posts, those supplied by the host organisation would remain in the employ of the host. At first, direct secondment was not ruled out nor the employment of people from outside. Nevertheless it was accepted that to ensure effective management, the staff from whatever source would be under the direction of the Head of the Project and would be required to meld into a homogeneous team. When with JET, loyalties would be to JET.

Rather quickly in the discussions it was appreciated that a multiplicity of background employers would pose problems and even harmonising the employment conditions of staff from just two employers would be difficult enough. Consequently the situation polarised to there being essentially two employers only of JET staff, the Communities and the host. Why did there have to be two? The smaller countries would have accepted, even welcomed, the scientific complement being temporary Euratom staff, whatever their source. At the Karlsruhe centre which hosts the Transuranium Laboratory of the Communities' Joint Research Centre, the German members of the Laboratory are Euratom employees and no one has found this arrangement particularly uncomfortable. In Britain, all the staff of the European Medium-Range Weather Forecasting Centre in Shinfield are on an equal footing and this is considered entirely normal. The organisation is independent of the CEC, but was set up in 1973 as a COST initiative. Staff contracts are for fixed terms but renewable and salaries are based on the coordinated scales of the international organisations.

Despite the precedents, the UK Atomic Energy Authority was adamant that as JET was to be built as an extension to an existing laboratory making use of a broad range of services provided by the host, it was vital that there should be complete parity between the nationals of the host regardless of whether they were working on JET, for JET, or otherwise. One factor in the discussion was, of course, the big difference in salary levels in Europe, especially between the UK and the rest. A factor of two going by normal exchange rates is not unusual and although it can be argued that the cost of living is lower in the UK and the social benefits marginally better (sic), these would also be to the advantage of the non-British staff. The result has been a big disparity between the employment terms of two people doing the same job in JET when one comes from the host country and the other through Euratom.

Had JET been built on a site isolated from any other station the argument might have been more evenly balanced, but it is likely that the Authority would still have insisted that if the British staff were seconded, they must conform to the Authority's standard employment conditions. Moreover other Associates would have been opposed to the idea of all people working for JET, even those in service positions, however menial, benefiting from Euratom employment conditions. France in particular, feared another Ispra although it does not object to all the staff of CERN, half of whom are French (weighted strongly towards the lower grades), with a high proportion living in France, enjoying the high salary levels paid there. CERN, it is argued however, is a permanent institution with only a few staff members that have re-integration rights (Wüster was one of the exceptions). Some consideration was given during the discussions to adopting the procedures used at Dragon. There the staff continued to receive their salaries and mission allowances from the home organisation which was then reimbursed by the project according to the post grading (which followed that of the Authority), but nobody liked the system or thought it appropriate to a Communities' venture. The Commission would certainly not have accepted this as a solution.

Most countries saw the options as equally acceptable and in view of the strength of feeling shown by the British were prepared to accept the two employer solution although it was far from evident that this is a better or fairer option. Some were tempted to believe that the Authority's attitude stemmed from a British posture of dividing the world into us and them or simply a churlish objection to any of their employees getting a better deal. Large national institutions as old as the Authority (formed in August 1954) have a reputation for inflexibility and fusion was rather far down on its list of priorities and so could not expect any favoured treatment.

Neither of these theses can, however, be taken too seriously. The Authority (in particular in the persons of Arnold Allen and Pease) took the structure of JET in all its facets with great seriousness and regarded the well-being of the staff as of prime importance. They could be accused of an error of judgement, but not of casualness or malice. The conviction was total that if JET came to Culham the Authority staff should be treated as one. Nor was it conditioned by the hope, still current when the principles were being discussed, that the building of JET would be contracted out to the host. Long after all such questions had been settled, the Authority (in 1981) appointed W. 'Mick' Lomer as Director for the Culham Laboratory to take over from Pease. He asserts that he would not have accepted the position under any other conditions. And he was already experienced in international laboratory management, having come from being Associate Director of ILL in Grenoble. There the staff receives a salary that is the local rate of CENG plus 10%, and he agrees that this works well. Apply the same principle to Culham, however, and he believes that not a single continental scientist of value would work there because of the pay differential.

An echo of this effect was heard in regard to the teachers' salaries in the European School which had been fixed by the Board of Governors at a level lower that at other schools even though still above British norms. The non-British considered themselves harshly treated and the Netherlands Government was sufficiently moved to change the status of the Dutch teachers to give them a supplementary income.

If Culham people working in JET had been paid so much more than those merely working for JET (or on other Culham projects) Lomer argues that his laboratory would be unmanageable. Moreover such a system works against the long-term interests of the staff. They can be cycled so much more easily under the present system and be re-inserted into the Authority hierarchy when the time comes for their promotion. Were they receiving Euratom salaries they would resist any change and, as happened in Dragon, would progressively lose contact with their home organisation. Only later was the question to be raised whether under Community law it was legal to make a distinction between staff working in a Joint Undertaking on the basis of their nationality alone.

When the Interim JET Council through its Working Party came to consider the staff conditions of service, they were not being required to write a set of Staff Rules and Regulations in the normal sense of the term. These existed in all their detail for both temporary Community servants and employees of the Authority. What was needed was a set of 'Supplementary Rules' as they came to be called concerning the assignment and management of the staff, consistent with the broad principles set out in the Statutes. In addition to the points already made above, these established that the composition of the Project team would strike a reasonable balance between the need to guarantee the Community nature of the Project and the need to give the Director the widest possible authority in the matter of staff selection. The Members of the Joint Undertaking were to make qualified staff available and each Member having a contract of Association with Euratom would undertake to re-employ such staff once its work had been completed.

The distinction in the terminology is significant because for a country like Ireland with no fusion laboratory as such, undertaking to re-employ an Irish scientist or engineer posed serious problems, yet it was obviously undesirable to exclude the possibility of their working on the Project. The return ticket, however, was regarded as having so much importance, not least by Wüster, who attended all the meetings of the Working Group, that it was felt vital to have explicit coverage of such a situation. Finding a formulation within the Supplementary Rules nevertheless proved to be very difficult and finally it was agreed that there should be a formal entry in the minutes of the IJC to the effect that the prospects of external re-employment at the end of service should be taken into account when selecting candidates and when no assurance of re-employment could he given, only a fixed period contract could be offered.

For the Project, the arrangement offered considerable advantages. By not being an employer as such it was able to stand apart from any general negotiations regarding pay or conditions of service, it was remote from the trade-unions in the UK and from the arguments in Brussels. Any problems that were not specific to JET could be passed on. In addition the return ticket or limited period contract removed any moral responsibility from the Project to take care of the long-term careers of the staff and the work force could be adjusted in numbers and in particular skills to suit the tasks in hand.

The great weakness from the Project's point of view was that the area in which it could recruit was largely restricted to the Associations and whilst the Commisariat a l'Energie Atomique for example, constituted, in principle, a large pool of effort it did not have good people waiting to take up a job of limited duration. For the Associations as such, JET represented an increase in the European fusion programme of some 25% and they certainly did not have 25% of their staff going spare. Moreover JET was a new departure in the world of fusion, demanding a much greater engineering input than previous machines and the scientific engineers that JET needed were few and far between. Already during the Design Phase, Rebut had been permanently under-staffed and recruitment was to remain a major headache throughout the Construction Phase and beyond.

A further weakness was that the rule applied equally to people whose jobs were not at all tied to fusion — secretaries for example, and return tickets had to be contrived to avoid getting rid of someone just when they were becoming their most useful.

The Project could in the light of experience claim that, apart from this problem of finding enough of the right people, the system worked well. All those recruited into Euratom posts for the Construction Phase came with return tickets and the

21 who went back to their employer at the end of the Construction Phase (as well as the 20 others who had left the Project at their own request) had all found an employment that conformed with their guarantee terms.

The claim, however, rings rather hollow when it is appreciated that it is only a minority of staff that is recycled after a short period and the guarantee is, in any case, only to re-insert at the original level which, for a person who is in his period of maximum climb on the career ladder, is not at all attractive. Moreover when it became necessary to recruit staff into the Associations in order to send them on to JET, their return ticket in many instances was for a few months only and was solely a device to give the person time to look for another job—helpful but not a career guarantee. In practice, the legalities may have been observed, but not everyone was happy.

At least in regard to career advancement, the Authority employee is in a favourable position, as the Supplementary Rules were so written as explicitly to require the Authority to grant temporary promotion, where appropriate, to staff nominated by the Director of JET. A caveat is that this should not put them into a level more than one grade higher than their substantive (established) post, and whereas this might sound like a discouragement to promotion, through the cooperation that has been developed, it has meant that the substantive post does not lag too far behind the real responsibilities or skills of the person concerned. For the Euratom employee, the Director is required to inform the parent organisation of any promotion, but there is no continuing obligation on that organisation to change the employee's status.

Contracts for employment with the Communities evoke a picture of gentle affluence for life, but there are different types of contract and because the type offered to the JET staff was for temporary employees occupying a temporary post, (Art. 2a)—the other side of the return ticket—no pension was (at that time) included. Severance pay was not ungenerous, amounting to 13.5% of the integrated basic salary received plus interest, but a fair proportion of this would have to be spent maintaining pension rights in the home country. In 1984, the rule was changed so that 2a) staff after 10 years became entitled to pensions in equality with 2d) temporary staff (occupying a permanent post) and the value of the severance pay for those employed for less than 10 years was increased by about 70%. As a consequence the pension rights and the severance pay became much more valuable than any promise of a return to a previous position. Nevertheless, the new system contains big inequities as in some countries only one state contributed pension can be paid for a given period. Thus a staff member leaving after 9.5 years might collect a reasonably handsome severance pay from JET, whereas, a colleague who leaves a year later gets nothing, except possibly a marginally increased pension on his retirement.

In 1978 however, the conditions of service were built around the notion of a relatively short stay in JET, followed by—what exactly?

Despite all the discussions in the Management Committee, and the Consultative Committee on Fusion, when all the principles governing employment were laid down, this was not really thought through. As the conditions of service emerged, they barely covered a scenario in which JET was a project of fixed term and, by clear implication, the end of the line. The underlying assumption was that everyone went quietly home to their own laboratories. No thought was given to the implications of the next phase when JET would be followed by a new project or the JET assembly rebuilt. Yet, in either case, the team which had been built up over the years would constitute the main fraction of the new work force—otherwise much of the investment would have been wasted. By 1978, some 30 people had already spent five years at Culham working on JET and those that would stay the 12 years initially allotted to the Project would have added up 17 years. This is hardly temporary employment in the normal sense of the term and if a European programme was to continue, it was only the beginning.

The conditions would have been more appropriate had JET been a device which once built would then be operated by a small staff mainly from the host country for the benefit of teams of visiting experimentalists. A big synchrotron radiation facility might just fit this description if the amount of machine development and on-site beam line and detector development were strictly limited. Even at CERN where great pains are taken to emphasise the primordial role of the visiting scientist, the 'permanent' staff outnumbers the visitors by about two to one. And a fusion device is very far from either. The machine is the experiment and the sort of people who build it are, in the main, similar to those needed to develop and operate it. The small number of staff going home in 1983 was to bear this out.

Much more ingenuity is required in setting up an international development undertaking than has so far been shown. It is equally absurd that employees of CERN, for example, after 30 years employment there still receive the same allowance, home leave and repatriation rights as they did at the beginning. Moreover many would happily exchange these for normal residential rights in the country where they have passed the majority of their working lives rather than being made to feel permanently temporary.

Again, it may be convenient for an international project to shelter behind the rules of a remote employer, but for the staff, this is unsatisfactory and members of the Design Team were quick to express their disquietude. Although they had members drafted into the Working Group, the staff representatives as such were only allowed to present one paper and were then excluded. Not that they would necessarily have been all that more perspicacious and, in part, their position was a little unreal. They saw themselves as having carried the JET torch through all the difficult years, they had the experience and they should be asked what they wanted. Instead they found themselves with no standing and no status— staff would be appointed only after the Joint Undertaking had come into being and then it would be under the authority of a Director they mostly had never seen.

In practice, within the limitations of the structure already agreed by the Council of Ministers, their interests were in good hands. The representatives of the different laboratories and Euratom were all anxious to see the Project manned by an enthusiastic team and the Director of Personnel within DGIX, Jeremy Baxter, came over to Culham especially to reassure the staff and explain what was going on. Wüster was vitally concerned both as Director designate and as a person committed to furthering the interests of his people as individuals. At the beginning of the discussions he was at pains to stress the inadequacy of the lump sum payment for temporary agents who might be employed for half their working lives, and urged the Commission to look for ways round this. However firm he was when he felt it necessary to apply the strict letter of the rules, he was never one to neglect a case of hardship or hide behind officialdom when responsibilities had to be shouldered. A really kindly man, he demanded the power over his staff in order to protect their interests as much as those of the Project. This he was given by the Interim JET Council with some hesitation at first, as some organisations would have liked

Members of the first JET Council. Descending and from left to right: C. Cunningham (IRE), C. Salvetti (I), H-O. Wüster, M. Frérotte (B), P-H. Rebut, D. Palumbo, (Euratom), J. Horowitz (F), R. Saison (Secretary); Front from left to right: J. Teillac (F), G. Holte (S), C. M. Braams (NL), P. G. Oates (Culham), L. Rey (S), A. M. Allen (UK), G. Schuster (Euratom), N. E. Busch (DK), Mrs Elbaek-Jorgensen (DK), R. Toschi (I), G. von Klitzing (FRG), R. Wienecke (FRG), J. Alex (L).

H-O. Wüster unveils the new panel at the entrance to the site before D. Palumbo, A. M. Allen and R. S. Pease.

to have had the Executive Committee, others the Partners, involved at least for promotions.

Secondment did not mean a temporary loan at the choice of the supplier. It was total transfer requiring complete loyalty to be given to the Project during the time the people worked for it, and the Director, with the exception of the most senior staff members, whose appointments required ratification by the Council, had complete rights of selection on the basis of their suitability. The sole reservation came from the clause in the Statutes urging a national balance in the composition. The Supplementary Rules would apply to all staff and also the non-industrial personnel provided by the Authority under the Support Agreement. The Authority insisted and Wüster readily agreed that industrial staff should remain within the Authority's management, but they were no less subject to direction from the Project when working for it.

Whereas drafting the Supplementary Rules was complicated because of the need all the time to steer between the Communities' and the Authority's Rules, little controversy was otherwise in evidence and with the able drafting notably of Joan Fox of the Authority's London Office and Wüster's care, progression towards agreement was smooth and rapid. Within the boundary conditions, neither the Project nor the Staff have much to complain of. It is the boundary conditions that should be questioned, and further thought given to such aspects as the differentiation between long and short-term and indefinite appointments, the transferability of pension rights on a European scale, the need to open recruitment to people who are not civil servants and do not need guaranteed employment for life instead of hiding them away in some grey category. Particularly for European projects designed to lead to industrialisation, an easy interchange with industry should also come into the calculations.

General Agreement

Already at its fourth meeting on 21 February 1978, the Interim JET Council was able unanimously to accept the draft Statutes, and confided to von Klitzing the task of preparing the final version without further reference. At its fifth meeting on 16 March the text of the full proposal that the Commission was to present to the Council of Ministers was exhaustively discussed and it was left to the Commission to formulate the final wording. The Financial Regulations were finalised sufficiently for the last draft to be approved by written procedure and the Host Support Agreement and Supplementary Staff Rules were almost as far advanced, although the UK was still uneasy about assuming responsibility for radioactive waste disposal once the Project ended. The sixth and final meeting was held on 19 April when these last two agreements were cleared. One final effort was made to involve the Executive Committee in the selection of non-senior staff, but Wüster supported notably by von Klitzing and Horowitz staved that off.

At that same meeting it was learned that the Energy and Research Commission of the European Parliament had already given its unanimous blessing to the Project.

Strictly the documents on finance, host support and staff were to be agreed by the JET Council once the Joint Undertaking came into being, but it was highly desirable that all aspects other than minor drafting questions should be cleared in time for the Ministers' decision. Teillac urged the IJC members to exert all the influence they could on the members of COREPER and try to have the proposals accepted as a point A.

And this is how it went. The Commission's dossier gained the whole-hearted support of the Communities' Scientific and Technical Committee, COREPER approved and on 30 May 1978 the Council of Ministers gave its assent without further debate.

In view of the early opposition to the notion of a Joint Undertaking on the grounds that the Statutes would have to go through Council, the eventual smooth passage should not go unremarked. No one tried to make capital out of the situation nor hold the Project up to ransom. This first European Joint Undertaking to be created ab initio was launched in a spirit of total unanimity. Nor should the amount of work that had been accomplished in the previous five months be minimized. Although the Groupe de Liaison, the Consultative Committee on Fusion, the Committee of Directors and especially the JET Management Committee had prepared the framework, and the decision of the Council in October 1977 had established fixed boundary conditions, a great deal of detail had still to be agreed, much of which involved questions of principle. All delegates pay tribute to Teillac's leadership, his ability to differentiate between matters of real concern and matters of mere form and he steered the IJC through the shoals of triviality at a steady pace.

Wüster's presence also was capital. In the first place he was there—the man who had to see it through. Whereas Rebut had been the undisputed inspiration of the machine design and leader of the Design Team, previous committees had been working in a partial void, with no site chosen and no one to ask the crucial explicit question of any administrative decision—Can I make this work? In the second, he quickly won the confidence of the delegates who recognised they had chosen a man of wide experience in management, sensitive to the preoccupations of different countries and totally straight in his dealings. But there were many others whose contributions were vital to keeping up the momentum—the Chairmen of the Working Groups, those responsible for drafting the documents, the Commission staff led by Schuster who had to clear them through the different Directorates and harmonise the different language versions—many people in many countries. A notable team effort of which Europe can be proud. And in the background Palumbo, the architect of European collaboration in fusion, seeing the fruits of his 20 years of patient planning.

The JET Joint Undertaking came formally into being on 1 June 1978 with as member organisations: Euratom, the Belgian State, the Commissariat a l'Energie Atomique of France, the Comitato Nazionale per l'Energia Nucleare* and the Consiglio Nazionale delle Ricerche of Italy, the Forsøgsanlaeg Risø of Denmark, the Grand Duchy of Luxembourg, Ireland, the Kernforschungsanlage, Jülich and the Max-Planck-Gesellschaft, Institut für Plasmaphysik of Germany, the National Swedish Board for Energy Source Development, the Stichting Voor Fundamenteel Onderzoek der Materie of the Netherlands, the United Kingdom Atomic Energy Authority. (The Swiss Confederation became a member in August 1979).

Five days later, Wüster accompanied by Palumbo, Allen and Pease and in the presence of the Design Team and members of the Culham laboratory unveiled a new entry sign at the gate of the Culham site announcing that this was the entry also to the Joint European Torus. Rebut was away at the time and some of the Team were heard to grumble that JET should have been on top. Otherwise it was a joyous occasion and a time for celebration.

---

*Transformed in March 1982 into the Comitato Nazionale per la Ricerca e per lo Sviluppo dell'Energia Nucleare e delle Energie Alternative—ENEA.

MANAGEMENT

Wüster's two main problems in the transition phase from negotiation to decision and design to building were to prise the Project away from the Associations without losing their support and take the Project firmly into his hands without disrupting the technical drive. The key to both lay in the internal management structure and the people who would fill the more senior posts. To the laboratories and the Design Team, Wüster was an outsider, no experience in fusion and no familiarity with the people in the field. When he presented his proposed structure to the Management Committee, criticisms came from all sides, but not in any consistent way. Different people had different ideas and different preoccupations, and for each suggestion there were as many against as in favour. Wüster's plan was for a Directorate composed for the time being of himself, the Deputy Director (Rebut) in charge of a combined Technical and Scientific Department, and a second Associate Director in charge of an Administration Department. Within the first Department would be seven Divisions which took into account the main sub-systems of JET, and within the Administration Department, three Service Groups. The Directorate and the leaders of the Divisions and Services would constitute a Project Board to advise the Director in the execution of his duties. An Associate Director in charge of a Scientific Department would be appointed in due course. IPP was strongly in favour of forming immediately a Scientific Department, but despite this and the lobbying for a tokamak physics division, questions on the directorate services and so on, Wüster bull-dozed his plan through and it was finally accepted by the IJC and became the framework for the Project.

Resistance also came to Wüster's contention that all posts should be announced, including those for which valid candidates existed in the Design Team. His argument was that senior appointments made by the JET Council should be more than a recognition of an existing situation. His reason was to make it clear that the Joint Undertaking started from zero and that no-one had acquired rights in it and no post (with the exception of Rebut) would be filled without his express or implicit authority. Although the majority of the Project Board and senior members of the Design Team, were evidently prime candidates for senior positions in the Joint Undertaking, someone had to be sacrificed to this principle and the unfortunate lamb was Eckhartt. His main qualification for the role was that he was German and his return to Garching after five years in Culham typifies the problems that a family can face in such circumstances. Neither of his children could be re-inserted in the Bavarian secondary school system — his daughter, aged 18, stayed on in England to complete her A levels, whilst his son, aged 14, tried for a time to reintegrate but finally came back to England to finish his schooling. The existence of the European school would no doubt have eased the situation, but would not have solved the problems. Children may adapt rapidly to the environment of an international school but this is not to say they can slip back easily into a state system.

Countries make great efforts to have 'their men' in key positions, but it is doubtful if Germany gained anything from Wüster's appointment. Far from showing favours to fellow nationals, he was always at his toughest with them. With the German delegates on the Executive Committee for example he was quicker to react than with delegates of other countries and he reserved his most violent outbursts for them. His early battles with von Gierke, normally the most courteous of men, but no light weight in form or argument, have been likened to collisions between a lion and a bull. In reverse, Wienecke was probably Wüster's least fervent supporter in the Council.

Wüster's autocratic style and his tendency to crush criticism in the Executive Committee with unnecessary force could have created a lot of hostility, but his political sense, his exceptional ability to recall and juggle with figures, and his general command of the situation were impressive. Moreover, his innate sense of fun, his remarkable grasp of English and French as well as German — not just the language as such but the literature and the culture — endeared him to almost all. He had the knack of remembering names and family histories, and he showed an evident concern for his people's well-being. His rumbustious manner created an aura of jollity that invaded the whole Project. To outsiders he would defend his staff to the limit, but in the face of what he saw as cheating, whether it was a minor sum in an expense account or distorted scientific reasoning, his rages could be frightening. He had the reputation of being a superb actor always putting on a show and calculating his losses of temper with impish delight. Such an interpretation loses sight of the emotional pressures that built up inside him when it was a matter about which he felt strongly. These he would sternly control until the moment for an explosion was opportune. His consumption of nervous energy was far above average and he felt the need to stoke his reserves at a corresponding rate. A long and difficult meeting could leave him totally exhausted, but he was at pains to hide the fact from those around him. Yet he could be patient and wait for his staff to do what was needed when he considered that precipitate action would be harmful in the longer term. And he could be calm and reflective. It was always a privilege to be with him when in one of these moods as one could fully appreciate the wise, cultured caring man that he was. He was always accessible, his desk was always clear, for his files were in his mind.

Vital to the success of the Project was the creation of an effective working relationship between the Director and his Deputy. It was a curious position for both. Wüster was taking the responsibility for the construction of a device whose design and costing had been done without him. He was coming in at a time when many of the major systems had already gone out to tender and in a number of cases contracts for stage 1 had already been let. A number of his contemporaries were surprised that he could accept such a responsibility without first instituting a design review and a costing check, although how this could have been engineered and who could have done it were not stated. Rebut made no secret of his resentment at Wüster's intrusion. Not just the broad design, but so much of the detail, even to methods of manufacture he felt, were his. He had kept the Design Team together during the difficult years and inspired them to keep going. What he wanted was to be allowed to get on with the job of building the thing he had conceived without interference from outsiders.

In view of the characters of the two men it could have been a recipe for permanent civil war. Although more restrained in manner, Rebut was, in his own style, every bit as autocratic as Wüster, as determined to have things done his way even though an alternative proposal might have worked just as well, as charismatic to his immediate collaborators, and just as impatient with incompetence. Whereas Wüster was cautious and basically analytical in his approach to policy, however, weighing up the long-term objectives and balancing the immediate gains, Rebut was bold, ready to trust his instincts and commit the Project to long-term decisions on the basis of his immediate appreciation. Where they met was in their sense of humour and in their common understanding of what JET had to do and their common determination to let nothing personal stand in the way of it being achieved. The result was a dynamic confrontation wherein Wüster came to appreciate and generously acknowledge the scientific genius of Rebut while he in his turn came to realise the special skills of Wüster and the fact that they were deployed in the interests of the Project as a whole. Disagreements there were, but the rifts were few and the staff was not required to choose sides.

In the documents approved by the Council of Ministers in October 1977 and then again in May 1978 a maximum staff of 320 was authorised of which 150 would be filled through temporary Euratom posts and the rest, by implication, by secondments from the host country. At the time of the site decision the Design Team comprised 34 people in all, even though some 80 persons had appeared on JET's books at some time and in R5, a complement of about 80 had been foreseen as being in post at the end of the Design Phase and the start of the Construction Phase. A vigorous recruiting campaign was necessary and even during the interim phase a number of people were selected for more senior positions including Jean-Yves Simon of the CEA to look after Personnel. These were taken on initially under the mobility provisions of the extended Design Phase Agreement as it was not until

1 November that the Commission delegated authority to the Director to sign contracts on behalf of Euratom. In the interim phase the precise gradings on the Euratom scale had still to be worked out and not all those chosen decided in the end to join. Small though their numbers were, the role of the members of the Design Team who stayed on was vital in maintaining continuity into the Construction Phase. Under the new organisation the majority were offered posts in one of the seven Divisions in the Scientific and Technical Department headed by Rebut. Gibson gave up some of his previous responsibilities to head a new Division of Experimental Systems that would in due course become the Scientific Department; Magnet Systems, including the mechanical structure were grouped together under Huguet, who thus became formally a member of the Project Board as he had been in effect throughout the Design Phase; Power Supplies continued under Bertolini, and Control and Data Acquisition, for the time being while the system was being defined, under Noll. Poffé was put in charge of the Assembly Division and these Division Heads were joined by Georg Duesing from IPP as Head of Plasma Systems (including additional heating) and Celio Vallone from CNEN in charge of Site and Buildings. Smart was made Deputy to Rebut with particular responsibility for supervising contract specifications.

Filling senior positions within the Divisions were other names familiar from the Design Phase. Amongst the Group Leaders were: Rudolph Polchen originally from IPP (Toroidal Field Coils and Cooling Systems); John Last, AEA (Poloidal Field Coils and Instrumentation); Etire Salpietro, CNEN (Mechanical Components and Assembly); Tulio Raimondi, CNEN (Engineering of Assembly and Remote Handling Group); David Booker, AEA (Planning and Organisation); Karl Selin, Sweden (Large Power Supplies); Pierre-Luigi Mondino, CNR (Advanced Power Supplies). They were joined by Franco Bombi, CNR who had been working part-time with the Project (Computer Group); and Gerhard Venus, IPP (Pumping and Gas Handling). Barry Green, an Australian ball of energy from the AEA and one of the original Team members who had spent so much time and effort over the years (though not alone) preparing the Project's various reports, became Rebut's personal scientific assistant.

A few names were missing: Sheffield decided that the opportunities for an English physicist in JET were too circumscribed and sought his fortunes in the USA. Pellegrini also left. The majority nevertheless remained, soon to be caught up in the rising tide of newcomers.

An urgent requirement was the appointment of the Associate Director for Administration as although the nucleus of the technical staff existed, the administration had to be built up from almost nothing. Even Manfred Bauer, who had come from IPP in 1974 and with one or two assistants had looked after all the administrative business of the Design Team, piloting through the staged contracts, had already accepted an appointment at Ispra and was only helping out at JET part-time. Having been a devoted servant of the Project over four years, he realised that as a German he could not become the new head of administration and he preferred to leave. JET was thus not a fully going concern by any means when the Joint Undertaking came into being and there was no army of recruits waiting to pour on to the Culham site the moment the agreements entered into force. It was rather a modest group that began life in the temporary huts erected on Culham's green open spaces and it was only a trickle of new people who came to swell the ranks in the ensuing months. A mere 100 posts out of the budgeted figure of 180 had been filled by the end of the year and only 3/4 of these were yet on site. Over 20% of those actually on the JET books were contracted draughtsmen. However, all but one of the technical Division heads was installed, George O'Hara from Ireland had been appointed Associate Director for Administration and two of his three Service Heads were in post. The essential framework was in place.

MAIN PARAMETERS OF JET

Plasma minor radius (horizontal) a		1.25 m
Plasma minor radius (vertical) b		2.10 m
Plasma major radius R		2.96 m
Plasma aspect ratio R/a		2.37
Plasma elongation ratio b/a		1.68
Flat top pulse length		20 s
Weight of the vacuum vessel		68 t
Weight of all the toroidal field coils		384
Weight of the iron core		2567 t

	Basic	Extended
Toroidal field coil power (peak on 13 s rise)	250 MW	380 MW
Total magnetic field at plasma centre	27.7 kG	34.5 kG
Plasma current — circular plasma — D-shape plasma	2.6 MA 2.8 MA	3.2 MA 4.8 MA
Volt-seconds available to drive plasma current	25 Vs	34 Vs
Additional heating power	4-10 MW	25 MW

THE JET DEVICE

Even though substantial developments had gone on over the years since R5 was first published and a number of the sub-systems had been totally transformed, the machine that was to be built was in its broad conception the one that had been presented to the Supervisory Board in May 1975. In the intervening time, more detailed analysis had shown where design improvements were needed or where tests should be carried out. Manufacturing problems had been exposed and special equipment developed.

JET as a Transformer

JET is an iron-cored transformer in which the secondary winding is a single turn of gas, the plasma. This plasma is surrounded by coils that produce a magnetic field in the same direction as the secondary current. In the transformer that used to be familiar in the power supplies of radio sets before transistors took over, the iron was usually in the shape of a double window frame, the two frames having a common central column around which the primary coil was wound. JET has eight frames each with a common centre and they give the machine its basic eight-fold symmetry with each octant being very like the others. Overall, the frames are 11.5 m high and although originally the radial dimension was estimated at 7.4m, when the design finalised in 1977 it had grown to 8.5 m and the section of the vertical limbs was $2 \times 1 \, m^2$ and that of the horizontal arms $1 \, m \times 1.4 \, m$ deep. The different pieces are made up from steel laminations 1.5mm thick (to take advantage of standard transformer practice) and clamped by insulated bolts between 75mm thick plates. Altogether nearly 2700 tons of steel is required. At full field, the magnetic forces pulling the whole structure together are equally impressive. Nearly 500 tons in the central column and some 200 tons between each of the horizontal arms and the outer vertical limbs. These are made in two pieces to limit the load on the crane when mounting.

The design of the centre section went through a number of alterations. In R2 a waisted column had been foreseen, but this had already given place in R5 to a straight column around which 10 coils were stacked on top of each other — the central poloidal field coils. In R8, giving the design changes up to the end of 1976,

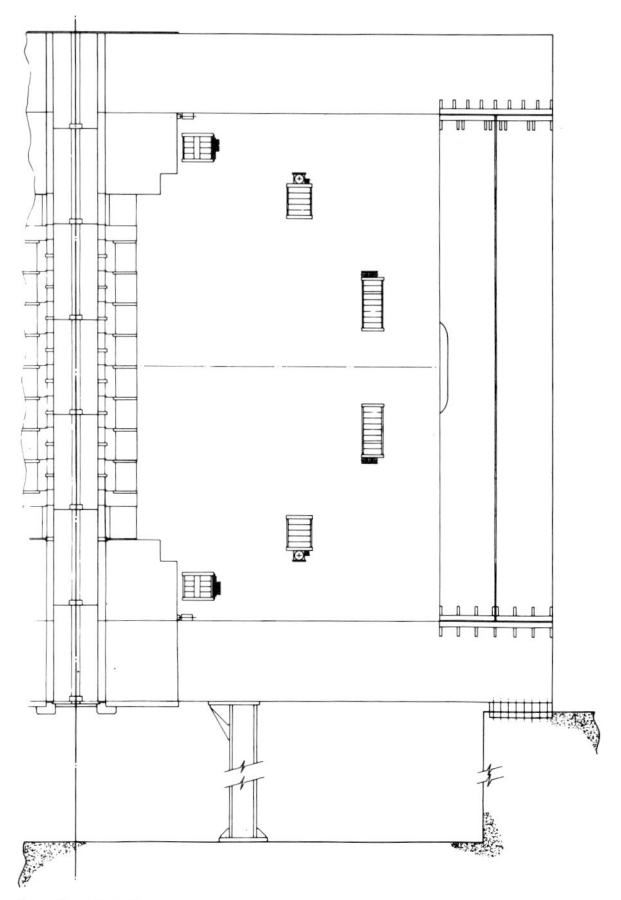

The system of poloidal field coils that generate the plasma and maintain its shape and position.

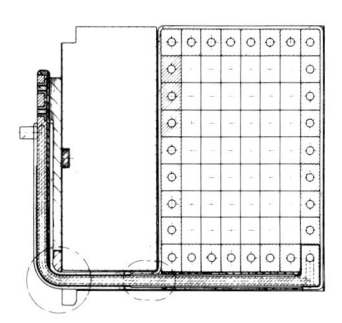

Section through one of the coils making up the primary winding.

12 coils were envisaged to give greater flexibility to the way they could be connected together, whereas the final design was based on eight coils of the same total length so that the pressure exerted by the toroidal field coils could be taken up on (six) complete units. The early decision of Rebut to absorb the radial pressure of the toroidal field coils on the poloidal field coils, whilst saving space, introduced difficult design problems, because in addition to squeezing in towards the middle, the toroidal field coils try to buckle, twisting one way above the plane of the plasma current and the other way below. Moreover, all the components are subject to expansion as the temperature rises when current is flowing so that totally rigid assemblies are excluded. In R5 already it was planned to seat the toroidal field coils in grooves machined in an intermediate hollow cylinder cut into eight similar sections, but this did not solve the problem of the interface material that had to withstand the high pulsed pressures and yet permit sliding at the same time. This was vital to the whole conception, but finally a PTFE-glass composite was proved to be satisfactory and able to stand up to high levels of radiation as well. The magnetic circuit is completed by a central core made up of insulated hexagonal rods around which is an annular space for the electrical and cooling connections to the inner poloidal field coils.

Each of these consists of nine concentric layers of hollow copper bar, eight turns deep, wound in a circular rather than helical fashion, the transition between turns and then between layers being made with a short double bends. The bar itself is about $54 \times 40 \text{mm}^2$ section with a circular hole of 15mm diameter for the water cooling. At one time it was feared that a rather more exotic cooling system would be necessary to limit the temperature rise but more refined calculations indicated that with the dimensions indicated, the rise during a pulse should not exceed 26°C. Insulation is Kapton and glass fibre. After winding, complete coils were given a final insulating wrap, mounted on steel support rings and impregnated with epoxy resin which was cured at high temperature so that on cooling they would shrink onto the rings. Provision was made for mounting all eight coils and connecting them electrically in two sets of four in parallel outside the machine so that they could be lowered as a unit into their allotted place.

One of the most troublesome problems was how to make the interface between the cylinder and the six coils taking the thrust of the toroidal field coils, whilst still providing for dimensional changes. Many ideas were explored and materials tested and only at a late stage was it decided to have the complete coil stack machined on a vertical lathe after assembly and to use as interface material the same PTFE-glass composite that was employed outside the cylinder. Because of the high pressures the windings must withstand and the need for great precision, the inner poloidal field coil system is far from simple.

The Toroidal Field Coils

The toroidal field coils are also far from conventional objects — very big, D-shaped, tapered and of thinner section on the inside in order to be able to crowd them into the middle as tightly as possible, and requiring cooling during the current pulse. It should be noted that JET was the first toroidal device to be planned for pulse lengths of tens of seconds and this introduced a totally new requirement in both the magnetic and power supply systems. The currents flowing in the different coils are not that much bigger than had been produced in smaller machines, but they last for very much longer times. JET has to supply and dissipate energy packets an order of magnitude higher than previously experienced. Each of JET's 32 toroidal field coils weighs 12 tons, and is made up of two pancakes each of 12 turns, the turns being locked together by an insulated key to take up the shear stresses between each turn. The overall dimensions of 5.68 m × 3.86 m remained unchanged subsequent to their definition in R5. Small changes were made to the disposition of the cooling channels, but otherwise attention was paid first to getting high quality bars drawn of sufficient length, then making detailed stress calculations, fatigue testing assembled sections and sorting out the coil manufacturing problems.

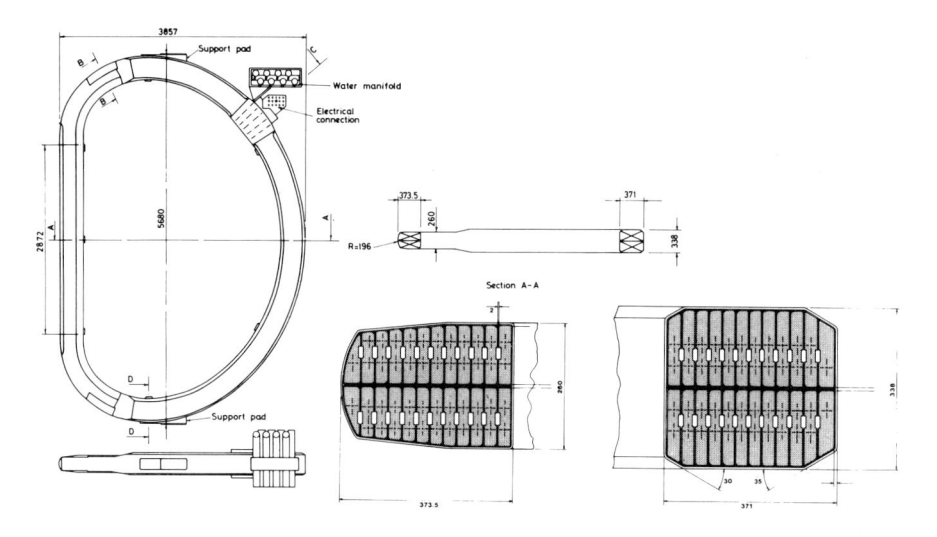

Elevation and Section of a toroidal field coil.

Two 4-ton lots of copper bars were ordered in July 1975 from which one was chosen despite its higher price because of its being more consistent in quality. The following year a contract for 40 tons was placed and in April 1976 the coil manufacturing contract was awarded. The maximum cross-section of the doubly hollow bar from which the coils are wound is over 40 cm^2 and the butt brazing of such bars (not available in lengths much greater than a single turn) proved to be far more difficult than might have been supposed. Brazing copper pieces together is one of the first exercises an apprentice metal worker will be shown, yet at these dimensions and with no sleeving of the water channels, standard techniques failed to produce a perfect join and the JET team was heavily engaged with the manufacturers in developing a reliable method. The key was found to lie not in magic chemicals, but in high cleanliness coupled to uniform heat distribution and a homogeneous brazing sheet. Over 100 experimental joints were made and then subjected to rupture or dynamic tests and a stringent testing sequence for each production joint (of which there are about 1500 all told) was elaborated as a result.

The first winding tests were begun in September 1977 (using cardboard instead of the proper inter-turn insulation) and from measurements of the resulting shape, modifications were made to the form of the winding mandrel to compensate distortions. A complete lower pancake was finished in April 1978 and using this for a second iteration it was planned to begin winding the prototype coil in June, straight after the Joint Undertaking came into existence.

Mechanical Structure

Rebut's determination not to encase the toroidal coils in individual steel boxes — impossible if he were to attain the high aspect ratio he was intent on, meant that

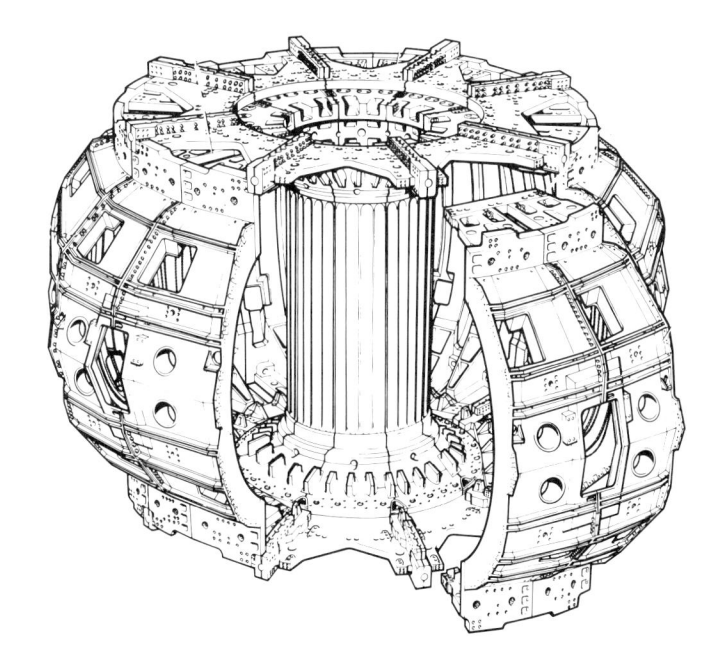

A cut-away view of the mechanical structure.

the coils had to be held in a strong framework that would withstand the huge twisting forces to which the coils were subject. That framework would also have to contain and support the vacuum vessel in which the plasma is formed and also allow access to it in many places for the additional heating injectors as well as all the diagnostic probes. One further limitation was that insulating breaks had to be provided particularly in meridian planes to stop large eddy currents circulating in the structure.

Rebut proposed, already in R2, fabricating a roughly spherical shell completed by top and bottom rings and pierced by a central cylinder, the outer shell wall providing channels into which the toroidal coils would fit. Octants would be built up with four coils and one eighth of the vacuum vessel already in place and would then be slid into position between the transformer limbs and attached to the neighbouring octants. In essence the general concept remained unchanged although the actual design was subject to major revisions during 1976 and further developed during the following year, partly because of the new ideas to operate with increased plasma pressure that could only be contained by raising the poloidal field and so the stresses on the structure.

Initially, the design envisaged the shell being made of light alloy sections bolted together, but as the performance targets were raised, it became clear that to have the required overall strength a welded construction would be preferable. However, when more detailed calculations were made of the stresses in the corners of the main penetrations, it was appreciated that the thickness of light alloy required was too great and a ferrous material would have to be used. Accordingly study contracts were placed with industry to determine whether it would be practicable to cast major sections in austenitic iron or austenitic stainless steel, or whether it would be desirable to build up a welded stainless steel structure from rolled plate and bars. Of the three possibilities, the last was certainly the most sure, but also the most costly in terms of time and labour, and the decision was taken rather rapidly to go for castings. More experience was available in industry of casting large components in austenitic stainless steel and rather surprisingly perhaps, there was little data available on the sort of imperfections to be expected with austenitic iron. Nevertheless cast iron has a higher yield strength, better casting properties and twice the electrical resistivity of stainless steel and would cost much less. This decided Rebut, despite the nervousness of his staff, to go for iron provided the tests commissioned by JET indicated that the material would be good enough. To be sure however, a prototype of the most complex piece was ordered in the Spring of 1978 so that final tests could be made. Happily, these were satisfactory or delays in the programme would have been inevitable.

A far from trivial matter in the design of the mechanical structure was providing for electrical insulation between the different sections — latterly in the equatorial plane as well as in the meridian — yet bolting them together firmly enough to withstand the twisting forces that makes each section want to slide over its neighbour.

This is prevented by inserting insulated keys, that take up the strain and a programme of testing was needed to prove that these and the metal round the keyways also would indeed take the load. In the process a circular section was preferred to the hexagonal section originally considered. Insulation has also been provided between the outer shell and the upper and lower rings which have to withstand big torsional forces. Far from being simple end plates, the rings that close the structure are very complicated pieces. Each ring is made up of two parts, the inner part being a conical shaped collar provided with retractable teeth that locate the toroidal coils and absorb the sideways forces. Relative movement between the collar and inner cylinder sections is taken up by dowels sliding in low friction spherical bearings. By the time it came to be built, the 460 ton mechanical structure had not only absorbed a very great deal of design effort, it had also given rise to a vast amount of stress analysis and model testing to confirm the validity of the ideas.

Vacuum Vessel

Similarly for the vacuum vessel, the general form of which was little modified subsequent to the decision between R2 and R5 to make it of double skinned construction throughout. A doughnut with a D-shaped cross-section enclosing a volume of almost 200 m^3, it is required to be able to hold a vacuum of 10^{-10} torr (considered 10 years prior to this as achievable only in small vessels under laboratory conditions) and at the same time to withstand baking to 500°C in order to

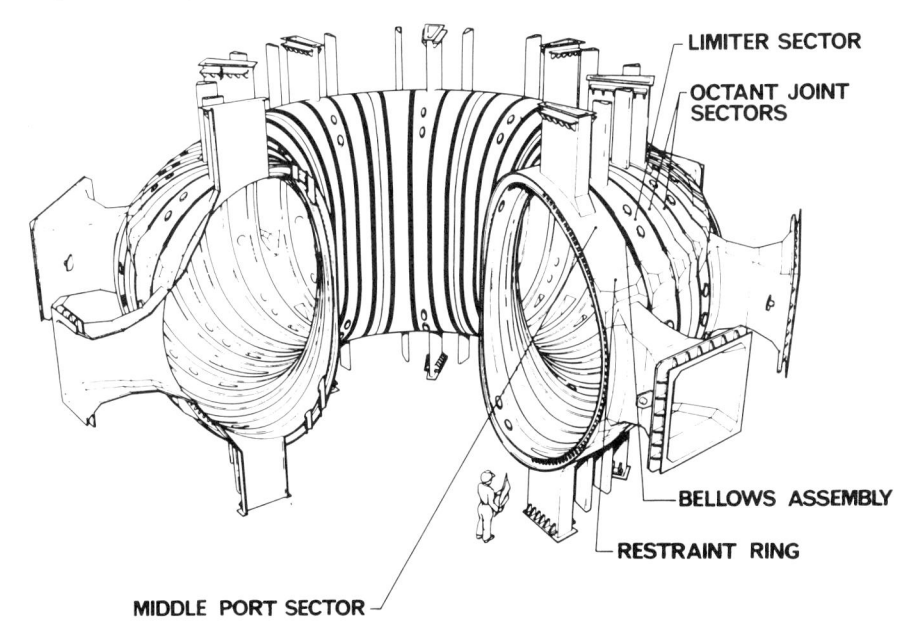

A cut-away view of the toroidal vacuum chamber.

drive out unwanted gases from the surface. Very much on the critical path for construction, an early definition of the design was important in order that the necessary testing work could be undertaken. Minor modifications were made after R-5 to the detailed shape, the limb of the D was given a curvature of 6 m radius to increase its strength, the distance pieces between walls were changed from columns to perforated webs, and the thickness of the material around the circumference was adjusted to equalize deflections under stress. The method of support — it weighs 70 tons — was also substantially modified.

From a mechanical point of view it would have been possible to envisage a completely rigid construction but such a form would have far too low an electrical resistance and there is no way of inserting insulating spacers whilst still retaining the necessary high temperature and high vacuum qualities. The solution was to include between rigid sections, bellows units whose wall thickness and so conductivity is about 1/10 that of the main vessel walls and whose convolutions quadruple the mean conducting path length.

An octant includes five open D-shaped boxes with walls of 12-20mm thickness and webs of 8-10mm thickness which take up all the mechanical stresses. It comprises a central rigid box with big rectangular access ports above, below and radial; two smaller boxes with radial openings for the supporting mechanisms for the movable water-cooled limiters (which scrape the edge of the plasma); two joint units on the outside. In between each box is a double bellows unit. All nine pieces are welded together. The spaces between the walls within an octant are interconnected so that either the volume can be pumped down to vacuum as a back-up to the main chamber or, it can be coupled to a supply of hot gas to raise the temperature for out-gassing or even, when operating at high power and the inner wall becomes hot, to a gas cooling system. Making the horizontal ports double-walled would have cost too much space, so these are single-walled and heated by electric mantles.

The need to be able to repair and maintain the system remotely when it becomes radioactive imposes severe restrictions on the design and the policy adopted was to make each octant a replaceable unit. Consequently the joint between octants demanded special attention. The main seal between adjacent joint units is between two flexible U-shaped rings which are fillet-welded together by an automatic machine, specially developed for the purpose, which crawls around the inner circumference. Rigidity is provided by plates bolted across the joint inside the weld. If an octant has to be removed the plates are unbolted, an automatic crawling device cuts the weld and with the new octant in place the weld can be remade. Inside the vessel, sensitive places like the bellows units are protected by plates against damage from the plasma. Outside, the vessel has to be thermally insulated to conserve the heat but especially protect the toroidal field coils.

But the really important quality of a vacuum chamber is that it holds a vacuum and the design called for some eight kilometers of weld run in a high nickel stainless steel, chosen for its high electrical resistivity and high strength at elevated temperature. There were those who doubted whether it would ever be possible to make the system leak-tight and it was clearly necessary to test out the ideas as early as possible. Already in January 1976, the contract for the manufacture of the first bellows section was released, a full-scale section of a re-weldable joint was manufactured in that year and the contract for the rigid sections and the assembly of octants was well advanced. At the first meeting of the IJC, agreement was given to order a full-size prototype rigid sector so that by the time the structures of the Joint Undertaking were agreed, most of the components of the vacuum vessel were entering their last phase of testing.

Shape and Position Control

While the inner poloidal field coils are the means for establishing the plasma and the current flowing through it, and the toroidal field coils generate the field that stabilises it, a further set of coils outside the structure, but inside the transformer limbs is needed to shape the plasma and maintain it at the right position both radially and vertically. These are the outer poloidal field coils and are mainly remarkable for their size, the two bigger pairs being nearly 8m and 11m diameter respectively. There are three pairs altogether, the innermost (coils No. 2) only being located clear of the mechanical structure. Consequently all three of the lower coils must be ready before the octants are moved in, and coils 3 and 4 have to be dropped down out of the way during assembly. If an octant has to be changed, the upper ones must also be jacked clear to let the octant be withdrawn. While coils No. 2 could be made up at the manufacturers and transported complete to JET, coils Nos. 3 and 4 had to be made in semi-circular sections and assembled at JET. In terms of design, the outer poloidal field coils are relatively conventional and after R5 only minor changes were made to fabrication procedures, although before they were finally ordered, the composition of each coil was made much more complex to give different possibilities of operation. Each of the coils No. 2 now looks like two coils one inside the other, the inner having conductors twice the section of the outer. A coil No. 3 consists of one pancake of 10 turns of hollow copper bar $82 \times 35 \text{ mm}^2$ followed by four double pancakes each of 10 turns of bar 35 mm^2. A single pancake is made up from bars cut to length and bent into a semi-circle the ends being turned out at right angles. Insulation is provided by wrapping in glass and polyimide tape. Double pancakes are made of two singles with a final common wrapping. Vacuum impregnation with epoxy resin completes the process. When assembled at JET, the different pancakes were stacked with glue between, and with the clamped electrical connections spaced at 45° from one layer to the next. The whole assembly was then clamped up and the glue cured. The most sensitive components in the system are the electrical and water connection units between the halves and already in 1976 a model section was manufactured and given over to fatigue testing. Copper too was ordered that year on a staged contract so that there would be no hold-up in supply when the time came to build. Altogether the six coils weigh 270 tons and

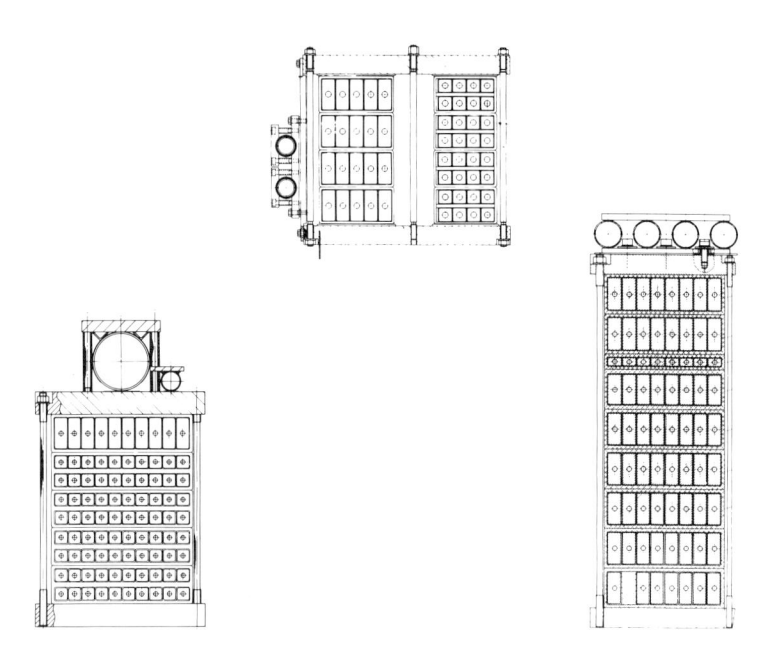

Section through poloidal field coils Nos. 2 (upper) 3 (left) and 4 (right).

they are subject to forces when JET is in operation of a similar order of magnitude. The upper coils are slung from the transformer arms, the lower are supported from the machine base.

Power Supplies

Above are described the principal components of the basic toroidal machine, but without power and cooling the device is inert. JET as we have already noted was designed to have an instantaneous power consumption only two to three times higher than some of the other fusion machines, but it was designed to have a much longer pulse length than previously attempted — tens of seconds instead of one or two. Consequently the total energy expended in a single pulse would be more than an order of magnitude greater than in previous machines. Already in R5 an overall peak power of 800 MW DC was anticipated and an energy consumption per pulse of up to 10,000 MJ (equivalent to 3000 kWh packed into the pulse period). With a repetition rate of once every 15 minutes, this means an average power consumption of only 12,000 kW and previous practice had been to rely on motor generator sets fitted with heavy fly-wheels which store energy in the rotating mass. Studies of the various possibilities of providing the peak power for JET soon showed, however, that this solution on its own would be very costly to install and would, moreover, to quote Bertolini, 'turn JET primarily into a generating station, and this was not its business'. He himself strongly favoured static systems, whereby the power is taken directly from the grid network through transformers and rectifiers. His continued contacts with CERN had kept him in touch with the new system that was being installed to power the SPS there and he was impressed not only by the relative cheapness of the system, but also by its inherent reliability once commissioned. Moreover, although an electricity supplier will charge more for a consumption that is pulsed, this is offset by the greater conversion efficiency of 70-80% as against the 50% of motor generators. Few grid systems, however, could and none wanted to cope with a pulsed load of the magnitude required — equivalent to a town of a quarter of a million people, all switching on at once and then off again after 20 seconds, to be repeated say 50-100 times a day. Both Ispra and Culham had very strong supply systems nearby (Culham being probably the strongest grid point of anywhere in the world) but Bertolini was constrained to devise a system that would fit any of the sites. A maximum pulsed load of 300 MW was originally defined in the site specification, although a somewhat higher figure was finally used as the basis for the design. This of necessity was a combined system that included rotating as well as static supplies.

By the end of 1976 with the poloidal field requirements much better defined (and with six months successful operation of the CERN SPS to give confidence) the principal characteristics of two systems, one providing 300 MW from stored energy and the other 450 MW as a pulse load on the mains, had been worked out and a call for tenders made for two flywheel generator convertors (FGC).

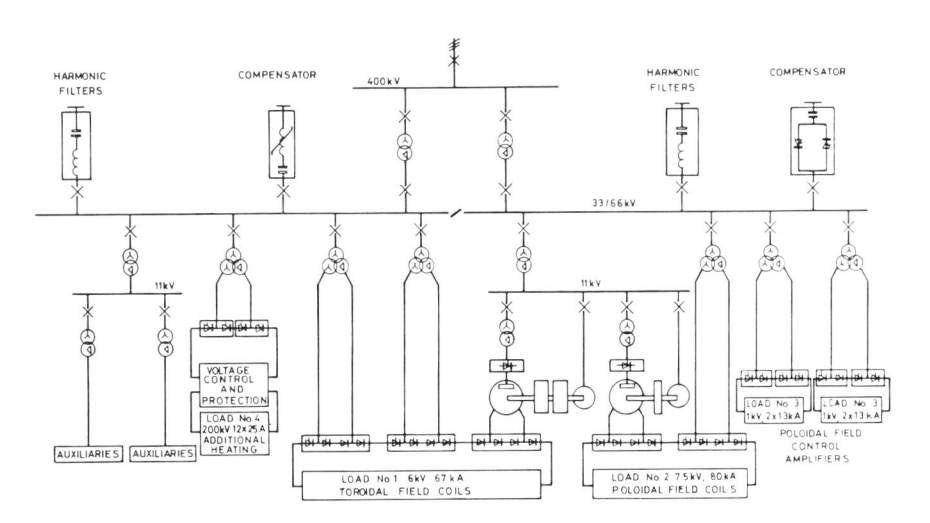

The JET main power supply scheme.

Two broad types of FGC can be identified, horizontal shaft machines that basically draw on steam turbine technology and which had been used on all tokamaks up to then, and vertical shaft machines that draw on water turbine and pumped storage experience. Many variants within these types had to be considered, but any solution would require an extension of established techniques and the JET team stipulated that it would need to satisfy itself that a full stress analysis had been performed.

In addition to the basic power feeds, there were numerous smaller supply systems to design and also analyse under fault conditions. Large quantities of power would be involved and a sudden short in a component must not be allowed to destroy a complete system—or endanger people. Altogether the major loads to be provided were for:

1. The 32 toroidal coils, all connected in series in which the current builds up typically over 13 seconds to 70,000 amperes (extended performance), and is then held steady during the pulse. Peak power is 380 MW.

2. The poloidal coils which drive and control the plasma. Only a changing field will keep the plasma current going so the coil current is increased in one direction over about a second—too slowly to initiate the plasma—and then the supply is cut which (as in a car ignition) creates the initial spark—the plasma—and the dying magnetic field starts driving the plasma current. This can then be maintained only by pushing more and more current into the coils in the opposite direction. In effect, because the loads are not simply resistive, there are two power peaks of some 300 MW separated by a second or two, followed by a more gentle ramp of say 20 to 50 MW.

3. The amplifiers which control the outer poloidal coils and which must be prepared even to reverse the current on some to keep the plasma in order. These take another 50 MW and more.

4. The additional heating which although still to be defined in detail needed a great deal of development. Quite apart from the inherent problems of providing high voltages in confined spaces, because of the liability to internal spark-over in such systems, the supplies need to be provided with protective switches capable of operating in times of microseconds.

Only when the site decision was known was it possible to enter into detailed studies of the interaction between the proposed JET power supplies and the UK network, but by the Autumn of 1977 the Team had completed the analysis of the offers for the supply of flywheel convertors received in April, and in February 1978 the JET Management Committee was asked to approve the placing of a contract for two identical low speed vertical shaft machines, one for the poloidal and the other for the toroidal system. The IJC endorsed the recommendation and the contract was signed in May 1978.

Control and Data Acquisition

Last of the major systems to be elaborated was the Control and Data Acquisition System (CODAS) and it was not until near the end of 1975 that sufficient effort was available for a special group to be formed to look after it. This was put under Peter Noll whose knowledge of tokamaks ensured that the system was developed as a tool for the scientists to use rather than just a piece of automation. Prior to that, the general principles of the system had been established and an estimation made of its size by a Sub-committee of the Data Acquisition Coordinating Committee — one of the specialist committees of the Associations answering to the Groupe de Liaison. They had concluded that the type of monitoring system that had been used previously on fusion devices where individual units were wired back to electronic boxes fitted with dials and regulating knobs was out of the question. The quantity of information that had to be recorded, analysed and stored, and the number of sub-systems with inter-connecting functions that had to be programmed and checked for the safe operation of the device pointed to the extensive use of computers. Moreover, as many measurements such as plasma position and shape were equally important for control and physics analysis, it was logical to plan an integrated system with inherent compatibility throughout. They were basing their judgement on the assumption that the diagnostic system keeping track of the evolution of the plasma during a pulse — not just average numbers for the density and temperature for example, but radial profiles of these and other quantities taken perhaps every tenth of a millisecond — would require up to 500 separate lines, each delivering some 200 readings per pulse *i.e.* 100,000 readings all told. For the control side they foresaw a similar number, notably during the commissioning period when, for example, some 3000 readings of the temperature around the vacuum chamber would be needed. Subsequent studies showed they had not been exaggerating. In R8 the objective is defined as being able to handle up to a million items of information a pulse.

This was not the only objective, however. Flexibility and ease of operation from a central point was high on the list, the ability to present immediately after a pulse the salient machine behaviour plotted in real time and also the plasma performance with calculations of such quantities as energy balance and confinement time. A further aim was to make it possible to commission different parts of the plant separately, then link the parts together so that they were subsequently accessible from the centre. CODAS was the system that probably gained the greatest benefit from the delay in the Project decision. There was time to analyse the system, study other systems and establish the standard patterns.

Experience at CERN was again of help as the SPS there was the first accelerator to be designed from the outset for centralised computer control — 10 km of machine, 6000 values to adjust and the margins for errors so small that a great deal of self regulation was indispensable. Despite the apparent similarity, there are nevertheless fundamental differences. The SPS is a quite rapid cycling (every few minutes) machine containing a very large number of identical elements spread over a long distance and designed to deliver a steady output over long periods. The experiments associated with the machine are totally independent and apart from synchronising signals between them, there is little interconnection. Moreover, the behaviour of the particle beams that are accelerated is fully understood and to a large extent, once tuned, the machine will run on its own.

JET though is both machine and experiment; quite elementary aspects of plasma behaviour are not understood and every discharge is unique. All big plasma producers have this characteristic; even without change to the set conditions of voltages and currents, pressures and injection rates, each pulse is special, differing from preceding pulses sometimes grossly, sometimes in subtle ways, but virtually never identical. Consequently a detailed record is needed for every one, not just for the moment but also for subsequent analysis perhaps months or even years in the future. Moreover, once a reasonably reproducible operating regime has been established, and its characteristics measured, it is time to change something and take another step in the experimental programme.

The JET CODAS is divided into two parts, one concerned with what is imposed *i.e.* control, and the other with what happens as seen by the diagnostics. Each of these is subdivided into smaller units with specific functions. A control sub-system might comprise a flywheel generator or a particular set of coils for example, a diagnostic sub-system a complete instrument. Whilst each sub-system is then further broken down into individual units, the two groups come together at the supervisory level making, in effect, a three tier structure. Communication between similar units is via the level above so that units talk to each other through the sub-system and sub-systems interact via the supervisory level.

At the supervisory level the threads are drawn together and finish in two interlinked (and interchangeable) consoles. One is primarily concerned with the machine control, checking that operator commands are acceptable and coordinating the control sub-systems, the other is mainly devoted to the experiments and the operation of the diagnostics sub-system. Each has its own computer. A third deals with the data coming back, making calculations, presenting them in readily comprehensible form, disposing of trivial material and then sending away for permanent storage in the big computer at Harwell, the essential information relating to each pulse.

The actual consoles which are similar to those used on the CERN SPS are fitted with just a few display screens, a keyboard, a knob, a tracking ball (for moving a cursor both vertically and horizontally) and a touch panel. Through these few devices an operator can set up a whole new group of parameters without ever leaving his chair. Through the touch panel he can call up the sub-system, see the breakdown of units, choose the one appropriate, see the sub-sections of that unit, choose the one he wants and so on up the tree until he arrives at the current or the voltage *etc.*, he wants to adjust. Or he can call in a routine and through a single command set a whole series of values that are interconnected. Extensive interlock systems prevent him doing anything potentially harmful and the security system signals back any abnormalities or departures from established limits.

Each sub-system has its own computer to which all the local units are connected via standardised 'CAMAC' modules which carry out routine jobs such as analogue to digital conversion, memory storage and recall, multiplexing (*i.e.* looking after a group of similar actuators, calling them up, identifying them, reading and setting or whatever is needed). A CAMAC 'highway' allows the computer to get in touch with any individual at a distance without interfering with the others.

For commissioning a particular sub-system, a portable or auxiliary console can be coupled in to the sub-system computer. Equipped with viewing screen, keyboard, joystick, knob and touch panel, it gives an engineer the means for checking out the sub-system close to where the components are located and it leaves the main console free for other activities. Once that sub-system has been proved out, the console is disconnected and the sub-system is linked directly into the appropriate supervisory level.

By the time the Joint Undertaking was formed, the whole CODAS system and the functions of the different computers (numbering some 20 in all) had been defined, and a preliminary enquiry had been sent out to manufacturers asking for their proposals. On the basis of the replies received a full invitation to tender would be prepared. In the meantime specifications for the CAMAC modules and the data highway were being prepared and the requirements for the interlock and safety circuits finalised. Had the decision to build JET been taken much earlier the number of people working on CODAS would have had to be greatly increased as a matter of urgency and the final result might well have been less coherent. As it was, experience regularly showed up requirements that had not been foreseen and the scale of the effort required to keep pace with innovations and modifications was continuously underestimated.

Buildings

Whilst Rebut had done his utmost to ensure that priority was given to those components that would be needed first or would take longest to produce, there was one crucial item that could not be tackled in real detail until the site decision had been taken. This was the complex of buildings that would house the machine, its power supplies and controls. Laboratories and offices were the responsibility of the host. Only in February 1978 was the consultant for the buildings appointed yet they figured on the critical path of the construction schedule. This proved to be one of the most frustrating contracts for both sides and the subsequent building contract generated more dissension and misunderstandings than any other. Together they showed how necessary it is in experimental projects for there to be close relations between client and contractor, and for the civil engineering to be treated with the same seriousness as the other disciplines. Compounding the inherent problem was the fluidity of the design. Whilst the machine itself and its major services were totally fixed, the definition of all the diagnostic equipment was still being done, support services such as cabling and pipe runs were still only sketchily worked out and the requirements for the operational phase had still to be evaluated. Moreover, the thickness of the shielding walls round the machine (some 2.5 m) and the massiveness of the foundation raft needed to support the machine, the 35 m square torus hall in which it is housed and the adjacent shielded cell and hot cells 35 m × 15 m — some 80,000 tons in all — meant that once poured, new penetrations would be almost impossible to provide.

Unwisely, the design consultant chose to maintain his drawing office in Newcastle which made for long communication lines and a tendency for detailed design work to be undertaken when the broad lay-out of an item was still in a state of flux. During this time new people were arriving on site, with new ideas and new demands which made for more changes than would have been the case if the site had been known early on and the buildings and the different systems had grown together.

It must be acknowledged also that the JET Team's own experience of big civil works was limited and the implications of demands for change or late decisions were not always understood. Vallone on whose broad shoulders the responsibility rested, started with a somewhat aggressive attitude towards the contractors and a state of general confrontation developed that was exacerbated by the language problem. Arguments over the exact meaning of phrases frayed nerves and wasted precious time and the atmosphere of relaxed cooperation that is normal in the UK between client, design consultant and contractor was soured in consequence. Steadily mounting costs, over and above the initial estimates, made for further strain on the Team members.

Had money been no concern, life would have been easier, but with a fixed budget, everyone was anxious to keep building costs to a minimum so that the maximum could be put into machine improvements. Cranage was, for example, limited to absolute essentials and clearances were kept to a bare minimum. An unusual requirement for such a construction, where the torus hall and the access and hot cells had to be isolated from the surroundings for radiation reasons, was to provide for the main crane to traverse from the Assembly Hall all the way through into the Torus Hall. Massive concrete doors open in the dividing walls and the whole upper sections can be lifted on jacks to let the beam of the crane pass underneath. The structures rising above the main building which house the jacking systems and the raised wall sections give the building its distinctive appearance. Where shielding is not necessary, as in the 35 m × 65 m Assembly Hall, a steel frame building with light alloy cladding had always been envisaged to give a big open area for minimum expense.

Wüster took a personal interest in the development of the building designs, both the specific buildings and the support accommodation provided by the AEA not only to see that costs were kept down, but also to see that the inhabited areas would be agreeable to work in and would contribute to giving a oneness of purpose to the staff. The entrance hall with its organogramme on which every staff member is identified by photograph and the space around the central meeting room for people to mix together during coffee breaks testify to this concern.

By June 1978 the soil investigations on site had been completed and the preliminary evaluations of what was required had been made. Not before the following year, however, could the main contracts be placed and buildings, it could be seen, would be the most likely cause of hold-ups in the years to come.

It was not just in terms of detailed design work that the machine itself was so much further advanced; already by the time the Joint Undertaking was formed, contracts had been placed covering about 75% of the fabrication. Altogether in the five years since Rebut and his advance Working Group had come to Culham some 186 man-years had been expended on the Project, counting also the consultants and support staff who had worked for more than a year on JET but not, of course, the people working under Articles 6 and 7 contracts.

Role of Contracted Work

Within the many committees that had been concerned with setting up the Joint Undertaking the role that industry should play had been a subject of continuous debate with many delegates favouring the placing of contracts for the development and manufacture of complete systems and even in the extreme case, sub-contracting the whole of the construction to a third party. Rebut had been unwavering in his determination to see that the control remained with JET and that JET took all the decisions up to the point where the task to be done was essentially manufacturing. Nevertheless this did not prevent him bringing in the expertise of industry at an early stage and confiding to industry specific developments or studies when the objective had been sufficiently closely identified. This applied to methods as well as hardware.

A number of delegates (whose experience of industry was perhaps limited to observation) doubted that JET would obtain the cooperation asked for when the results of studies would be open and no manufacturing commitment given, but the rules were clear and, in the main, companies welcomed the opportunity of investigating new technological situations. When it came to placing production contracts, the degree to which a system was broken down varied a great deal. Whereas a single contractor was engaged to build and install the two flywheel generator convertors, the sections of the outer shell structure were cast in Germany and machined in Italy — not from any desire to spread the suppliers, since interfacing between them added to the work of the JET staff, but because they were judged to offer the best terms. JET had no inhibitions about dividing the work as it saw fit, ignoring traditional commercial practices and national boundaries. It might be noted that the two companies put together by JET in this instance have since been tendering as partners for similar work in other fields. Again with the vacuum vessel, JET selected a British company for the octant assembly and for the manufacture of the rigid sections, using a nickel steel ordered in Germany and confided the bellows sections and U-joints to an Italian company using an Inconnel that came from the USA. Surface treatment of the inner wall went to another UK company that JET selected.

Once manufacturing began, Rebut was equally determined to see that critical items came under the close supervision of JET people. They were there not just to see that specifications were followed, but to help when difficulties arose. Not all companies were quite so ready to welcome this form of intervention, believing that they had the knowledge and there was little that outside scientists could contribute. Many, however, came to appreciate this collaboration and in a lot of instances, the more basic and original approach of the scientific mind produced solutions to novel problems when recourse to past experience failed.

The danger of such a centralised policy is that a design team spends a lot of effort developing products that already exist on the market or techniques that are entirely familiar ('re-invent the wheel'). Little of this can have taken place or the Team could not have got through so much work. Bearing in mind the time spent on committees and lobbying for a decision, staff changes and revising programmes, the total man-years was modest. A judicious use of study contracts was one explanation, another was the research mentality applied to problems— start by finding out what is known, and don't hesitate to learn from any source.

Contracts with industry were probably more effective than those with the Associations. Whereas between JET and industry there was a clear client-supplier relationship, this was less evident in the jobs that were done in the Associations. Explicit analyses, such as stress calculations on the vacuum vessel raised no difficulties, but when it came to the development of a sub-system, the Association tended to fit it into a broader research context. Although application to JET might be an essential element, it was not necessarily the dominating motive and control lay in the hands of the Association and not JET. This had an effect on time-tables and on compatibility with other features of the JET design. Such was the case in the development of the neutral injection heating system and some of the diagnostics (see Chapter XV), but sight should not be lost of the large numbers of smaller contracts that were placed with the Associations, essential to both the design and construction phases, and which were completed to schedule.

FORWARD PLANNING

Despite the depression that waiting for the JET decision had provoked, the mood of the fusion community in the second half of the seventies as a whole was buoyant. Construction of the TFTR at Princeton in the USA had begun in March 1976 and real plans were already being made for the next steps. In Europe too a small task force on Long Term Planning had been set up at the end of 1976 and had produced a report for the Groupe de Liaison and the Consultative Committee on Fusion on Long Term Planning Towards A Demonstration Fusion Reactor (the DEMO). In this report, published in June 1977, two further major devices were foreseen to come between JET and the DEMO, one to study all the problems of a burning thermonuclear plasma and the other to study non-nuclear questions, notably the technology of using superconducting magnets. The cost of these two was estimated at 3½ and 2½ times JET respectively, and the supporting programme 8½ times as much. The DEMO it was estimated would cost something over 6×JET. Moreover, acting on the assumption that JET would start up in 1981, the report strongly recommended that a DEMO definition group be established early in 1979 which would then continuously optimise the design in the light of progress and provide the information for planning the support programmes. With some adjustment to the time-table this became the blue-print for further action beyond JET.

A new stimulus to ambitions world wide came in April 1978 with a letter from S. P. Velikhov, Director of the Soviet Thermonuclear Programme, replying to a call for proposals issued by Sigvard Eklund, the Director General of the (UN) International Atomic Energy Agency (IAEA). He launched the idea that a joint planning group of some 15-20 persons be set up under the Agency's auspices with the task of developing internationally the design of a demonstration tokamak reactor. The proposal carried also the recommendation of A. H. Petros'yants, Chairman of the USSR State Committee on the Utilisation of Atomic Energy. In its preamble, it was stated that 'the thermonuclear programme of the USSR is directed towards the scientific demonstration of a controlled thermonuclear reaction by the mid 1980s. In particular, reactor plasma parameters are expected to be obtained in the

T-10 M tokamak*... the current position is that an engineering demonstration of controlled thermonuclear fusion is expected in the mid 1990s and final demonstration of the industrial generation of electricity by the year 2000'. Velikhov concluded by expressing the belief that it would be wise to begin to study the possibility of combining the efforts of the Member States of the Agency for later stages of the programme as well. In the preliminary phase, the Soviet Union was prepared to offer the widest possible cooperation.

Results coming in from operating tokamaks around the world were encouraging. Alcator in the USA and Pulsator at IPP had clearly shown that high density plasmas could be obtained with a low impurity content. Comparisons between the performances of different machines, notably FT at Frascati and PLT at Princeton, confirmed that confinement time indeed increased with density and also more than linearly with cross-sectional area. It could even be proportional to the area squared, a result which strongly favoured JET with its big vacuum chamber. Moreover, the D-shaped profile seemed not only to have been a wise choice from the point of view of limiting stresses, but also to favour the establishment of special profiles in which the pressure could be higher than in circular plasmas.

Above all, however, the injection of neutral particles subsequent to the early experiments begun on DITE in 1972, both into TFR in France and then ORMAK in the USA had been full of promise. Come the Summer of 1978, Princeton announced that with neutral injection they had reached some 80 million degrees in PLT even though the efficiency of conversion was less than hoped. This information first discussed at the 7th International Conference on Plasma Physics and Controlled Nuclear Fusion held that year in Innsbruck, and then in Paris in September at the 3rd IAEA Technical Committee Meeting on Large Tokamak Experiments put the tokamak world into a ferment, particularly in Europe. TFTR in the USA had been programmed from the beginning for some 15 MW of neutral injection and the Americans announced they were going back to double it at least. (The actual figure must be regarded with some caution as in European documents an initial provision of 24 MW is cited, the reason being that between the external power supplies and the final energy deposit in the plasma there are a number of stages of only modest efficiency and different laboratories used different definitions for the beam power, the tendency being to employ a high or low number depending on the thrust of the immediate argument).

Neutral injection for JET was still at the research stage and although in the submission to the Council of Ministers a figure of 10 MW had been included instead of the 3 MW (plus 1 MW of RF heating) written into the R5 estimate (the difference in cost coming out of the reserve fund) an efficiency of 40% had been assumed, double that which Princeton was now reporting and JET could not expect better.

---

*An extended version of a device that started operating in 1976.

JET had visions of TFTR coming into operation eighteen months ahead of them, switching on perhaps early in 1983 four to five times as much additional heating as was planned for JET and going to ignition while JET was still constructing a device whose basic performance would belong to a previous generation.

Hastily the pluriannual programme for 1979-1983 was modified to warn ministers that significant changes would probably be proposed when the 3-year review came round in 1981. Rebut explained to the Executive Committee the arguments for going straight to 25 MW additional heating and Wüster presented to the JET Council in November 1978 a revised programme for JET which had the target of advancing completion of construction from June 1983 up to the beginning of that year, doubling the power supplies for neutral injection up to 50 MW to take account of the expected lower efficiency, and starting in 1981 on the additions for the extended performance programme.

The reduction in construction time was to be brought about by cutting down the time between prototype delivery and the start of series production and compressing the already tight schedule for the construction of the JET buildings. The main items in the advance towards extended performance would be a further 15 MW of additional heating, possibly RF if development of the technology went well, a more powerful outer poloidal field system (for plasma profiling), and a speed up in the development of remote handling techniques for the early use of tritium.

That same month the Next European Tokamak (NET) group was set up and Toschi was nominated leader. Shortly after, in response to Velikhov's prompting, a study at first called UNITOR, later altered to INTOR, was initiated by the IAEA which grouped together (W) Europe, Japan, the USA and the USSR. NET's efforts were concentrated on coordinating the European contribution.

Competition across the Atlantic was to be a powerful spur (to both sides) throughout the Construction Phase of JET and it was a major bond in uniting the efforts of Wüster and Rebut in their different domains. As the fusion man who had been the driving force behind JET for the whole of the design phase, constantly trying to up-grade the machine's potential performance in one way or another, Rebut could not help but view the American programme with alarm even though he was convinced that TFTR was intrinsically inferior to JET in basic concept.

Wüster coming from CERN where the tradition was to build to time and within the budget forecast was determined to prove he could do as well. Moreover, his opposite number, Paul Reardon, at Princeton, was just the rugged friendly rival that Wüster loved to do battle with. He too was a high energy physics man having built the booster for the revolutionary (from the point of view of size and cost) 500 GeV accelerator at Fermilab and then taken charge of commissioning the whole machine at a time when it looked to Europeans as if the corners had been cut too close and the machine would never become operational. He, like Wüster, had a problem of acceptance by the fusion community and both took this as a challenge.

They were to have many a wager on the performance of their two laboratories over the years that were, in due time, honoured in splendid gastronomic style.

Not everyone embraced the competition with such fervour. Neither the Dutch nor the Swedes saw the need for the acceleration and were not really persuaded by Wüster's contention that it was really only a re-arrangement of the commitment budget that was involved. They feared that JET would begin to run away with the available funding and the first to suffer would be the poorer laboratories. Delegates to Council were in the uncomfortable position of being responsible for the well-being of JET and the promotion of its cause in their own country whilst being conscious of the negative impact that JET increases were likely to have on their own resources.

For the basic programme, their fears were largely unfounded and the construction figure of 184.6 MEUA at January 1977 prices was raised only modestly and even then in part because it proved impossible for the Associations to fulfil what had always been assumed to be their share of the programme. The 5-year plan that Wüster elaborated in February 1979, forecast for the JET device itself a cost that was lower even than had previously been quoted — 32.6 MEUA. As each month went by and new contracts were placed, further proof was obtained of the soundness of the original estimates for the main items and as a result the allowance for contingencies could be reduced. The principal additions concerned the increase from 3 to 10 MW for the neutral injection heating already agreed, the need to double up on the power supplies, a further 1.7 MEUA for control stations and 1.0 MEUA for remote handling equipment so that assembly crews could have experience of the techniques involved. The most disquieting modifications for the Dutch delegates, however, was the big increase in building area demanded, almost 50% increase in the Torus and Assembly Hall complex (a full test rig for neutral injection components was planned) double the size of generator hall, enclosures for electrical gear that previously was considered to be out-door and the addition of a separate control building. Altogether these raised the building costs from 13.8 to 16.6 MEUA and additional access roads added a further 1.6 MEUA. To set against these increases, the reserve was dropped from 15.2 to 4.2 MEUA (which the British and the French found disturbing) and 8.5 MEUA could be saved on staff costs. These had previously been calculated on the assumption that all employees would be paid Euratom rates. With JET located in England and 55% of the man-years being paid at AEA rates, an overall saving of some 20% could be foreseen. Such a transfer from personnel to equipment chapters was again not liked by the Dutch, but as the envelope was respected, the majority of the Council was prepared to accept the revised basic programme.

For the extended programme, a total of 62 MEUA was presaged to be spent over the years 1981-83 with no great precision on how the work should be phased, but to be sure that at least an extra 15 MW of additional heating and the toroidal field enhancement would be ready by the end of 1982.

At the same Council meeting in February 1979 the overall structuring of the Joint Undertaking was completed with Braams being appointed Chairman of the JET Scientific Council and Bickerton nominated as Scientific Associate Director heading the third Department for which delegates had pressed so strongly. This did not prevent them, however, the following month from cutting the staff posts from 320 to 300 the difference coming largely from that Department.

In subsequent five-yearly programme estimates which, at the request of the Executive Committee, Wüster produced every six months, the cost of the JET device and power supplies expressed in 1977 money continued to follow the estimates, whereas further adjustments had to be made to the ancillary costs. Cooling, cabling and buildings were the main culprits. In October 1979 a further 3 MEUA was added to the building costs and the figure for the pumping and cooling system was pushed up to over 8 MUEA from the original 3.5 MEUA. Because of lack of effort and its relative banality, only cursory attention had been given to the system during the design phase, and subsequent elaboration of the design was revealing it to be a more major item than had been appreciated. Reducing the contingencies accommodated the revision for the time being but when in the Spring of 1980 a further 3 MEUA had to be allotted to the pumping and cooling, another 2 MEUA for site work and 4 MEUA for assembly and maintenance, Council felt that the contingency was being reduced to a dangerously low level, particularly in view of the coming need to cope with the added load on the Project in the areas of both diagnostics and additional heating. Accordingly in June 1980 the overall budget for the construction phase was changed to 191.4 MEUA and again in 1982 (when all the contingency had been ear-marked) to 196.2 MEUA, finally to settle in 1983 at 198.8 MEUA.

By the Spring of 1980 however it had already become clear that JET would not be able to do everything by the end of 1982 and the target had been redefined to aim for completion of the main assembly by that time and to press on with commissioning with a view to producing the first plasma in June 1983. That would be difficult enough. Moreover, there was no chance of getting in 10 MW of neutral injection heating by then and although it was intended to install the equipment directly coupled to the vacuum chamber and some other components before switching JET on, it would not be practicable to install the beam lines before the end of 1983 at the earliest. Neutral injection for JET with its long plasma dwell time and need for high injection energy posed problems of a quite different scale from those that had been resolved on other tokamaks.

FUNDING

All the figures cited in the previous Chapter relating to the basic programme are at January 1977 prices, whereas budgets and payments had to be worked out in the currency of the time. When the problem of indexing had been raised in the Interim JET Council it had been decided that this should be left to the JET Council and accordingly Wüster prepared a proposal which, with a few modifications, Council was able to approve early in 1979. This set out the methodology whereby rises in prices and in salaries were recorded in a number of countries and combined according to a weighting factor that was related to GNP and probable supply level to JET, this last taking account in particular the special supply position of the host country. The figures used in the materials category were those presented under Item 620 of the General Index published by the Statistical Office of the European Communities in the Monthly General Statistics Bulletin. Rather than attempt to make the calculation for all countries a representative selection was used notably Belgium (initial weighting 4.5%), Denmark (2.5%), France (20.6%), Germany (29.4%), Ireland (0.5%), Italy (9.9%), UK (32.6%). The method was essentially retroactive in its application and as far as it went seemed to be reasonably balanced. Its application, however, entrained some anomalies and in the early accounts of the Project the payments budget was always larger than the commitments budget because escalation clauses within contracts appeared only in the budget when payment was made. Although this was partly taken care of from 1980 when the practice was instituted of making allowance in the commitments budget for contracts already placed, there was still no provision for the escalation that took place between the time of the last budget up-date and the date when payment actually fell due. Escalations also posed problems when comparing tenders and complex formulae were devised to try and cater for this. Even then, changes in exchange rates could still distort the situation.

When figures were being prepared for the 5-year pluriannual programme the sums presented were in the money of the beginning of the first year with no provision for subsequent inflation. This could lead to a really big erosion of purchasing power as time went on. The problem was exacerbated by the paradoxical position in the UK in the late seventies when inflation was running at up to 15% *per annum* yet the value of the pound was increasing with respect to a number of other European currencies. For the basic programme this was contained by having an agreed total figure but the application of the indexing procedure was not automatic and the first time it came to the test Council shied away. Assuming that 1978 would resemble 1977, calculations indicated that in the money of January 1979 the ongoing cost of the basic programme would be 211 MEUA and although this was corrected to 206 MEUA when the real figures for 1978 became known, Council reduced the sum to 200 MEUA by cutting the contingency allowance so that it looked better in the 1979-1983 pluriannual programme submitted by the Commission to the Council of Ministers.

This submission was running into real difficulties as the delegates to the Group for Atomic Questions adopted an attitude quite different from that of the leaders of the Associations. Controversy centred not on JET at all but on the overall volume of the programme that was being contemplated and especially on the proposals for inertial confinement experiments and a centre for tritium development. Two new tokamaks were under study in Europe—a medium sized superconducting machine in France and a large very high field machine, soon to be called Zephyr that was IPP's response to JET going to Culham. This last would be almost as costly as JET itself.

Although the Commission was not envisaging any substantial increase in the number of qualified professionals currently engaged on fusion (numbering 860 in mid 1978), it was putting forward a programme that at 1979 prices would cost some 736 MEUC—excluding JET. Simply continuing the programmes in the Associations was estimated to cost 588 MEUA at fixed prices over the five years (based on an expenditure of 104 MEUA in 1978), while extensions to the Associations' programmes were estimated at 103 MEUA, including 50 MEUA for the two new big tokamaks and 23 MEUA for additional heating developments outside JET. On top of this, new activities costing 45 MEUA were planned for tritium technology and materials. The contribution of the Communities to the total would be 217 MEUA (as compared to 124 MEUA approved by the Council for the period 1976-1980 and of which some 44 MEUA remained). In addition, the Communities would be paying their 80% share of the JET budget which for the five years was cited at around 250 MEUA. This was made up of the contrived 200 MEUA for the basic performance machine less the 12 MEUA paid in 1978, plus 62 MEUA for extended performance. The 25 MEUA that would be needed for operating expenses in the second half of 1983 were quietly ignored.

The programme had received solid support from the European Parliament, but once again the Commission found that the recommendations received from national delegations at one level were not sustained at another. Fusion still had its Groupe de Liaison and Committee of Directors in addition to the Consultative Committee on Fusion which was supposed to represent government thinking, but in effect there was little real selection between projects and recommendations for new projects proliferated. This was the reason for finally condensing the advisory system from December 1980 into a Consultative Committee of the Fusion Programme in which half the members were explicitly from the administrations of the different governments.

Meanwhile, in the USA, fusion had been receiving a great boost and confidence was running high. An Act passed by the Senate and House of Representatives in January 1980 provided for an acceleration of the current fusion program with a doubling of the funding at fixed prices within seven years, including a 25% increase in each of the fiscal years 1982 and 1983. Foremost amongst the targets was the achievement 'at the earliest practicable time, but not later than the year 1990, (of the) operation of a magnetic fusion engineering device based on the best available confinement concept'. In European terminology this was NET and the full DEMO was to be operational at the turn of the century.

In stark contrast to this governmental stimulus on the other side of the Atlantic, when the Pluriannual Programme for 1979-83 was finally agreed in March 1980, the Communities' support for the Associations suffered a 12.5% cut to 190.5 MEUA. JET, however, was given the support requested and its budget up to the beginning of operations was agreed at 201.25 MEUA. Unfortunately inflation was already eating into the value of this sum and when Wüster presented that same month his budget forecasts for 1981, he gave warning that unless new provisions were made, the commitments would become restricted before the end of that year. It was France that persuaded the Council then to face up to the problem and have the Commission forward to the Council of Ministers a real estimate of requirements up to the end of 1982 instead of scrambling from year to year. As a result of this the Ministers approved in March 1981 an interim up-date of the total cost to 263.75 MEUA at January 1981 prices. Despite this, delays in deciding the Commissions' annual budget for both 1981 and 1982 obliged the Project to operate within restricted budgets during the first half of each of these years. Whereas the JET Council could approve in principle the JET budget for a given year, it could only adopt the budget and thereby authorise the Director to enter into new expenditure, after it had received notification by the Commission that the relevant appropriation had been granted. The tardiness with which decisions are taken within the Communities has been a constant problem for project directors trying to follow a defined scientific programme. These delays waste time, a great deal of valuable management effort and generate a cynicism towards the Communities and their institutions that does permanent damage.

Although the practice of presenting pluriannual programme budgets in initial money without explicit provision for indexing remained in practice, JET improved its own annual forecasts by expressing them in terms of the prices expected to be in force in July of that year, — a procedure that the Commission's financial services found distinctly odd. In this way a better estimate was given of what the expenditure in real money terms would be and what commitments could be made.

This would be particularly important when JET moved into the Operational Phase and the basis of the budgets changed. During the Construction Phase and also for the enhancement programme, the Council and the Commission were agreeing to an explicit programme whose cost was measured against the prices valid at the time of the estimate. For the operational programme, the scale of activity was controlled by the annual budgets. When the different phases overlapped, care was taken to see that the different budgeting procedures were kept separate.

The accounting methods originally adopted by JET were modified in the light of comments made by the Court of Auditors. At their request, for example, from 1982 the commitments were adjusted to include outstanding legal liabilities. The Project, however, resisted pressure to replace the Income and Payment account by an Income and Expenditure statement, believing the former to present a truer picture of the financial position. It also resisted pressure to change its banking arrangements, and a joint examination with the Commission indeed led to the conclusion that there was no reason to do so. In particular, it was agreed that JET should not gamble with its current account even though currency changes could have an important impact on resources. With sterling outside the 'snake', the problem was exacerbated, but the Project had neither the skills nor the wish to play the market and Wüster believed it would be highly imprudent to contemplate any such activity.

PROJECT CONTROL

Wüster's overall financial control was strict and highly autocratic. It was based on two essential principles:

1. Firm control of the Project required direct control of the purse;
2. Responsibility for planning and supervising expenditure must lie with the scientists.

In the administration, the Project Control Office answering directly to Wüster played a key role and was much more concerned with budget control than the Finance Office, which came under O'Hara and whose functions were essentially related to accounting, ensuring that the financial regulations were observed and handling the actual money transactions.

Project cost estimates were worked out every six months initially with the Head of the Project Control Office holding direct discussions with the Heads of Division. These were not designed to be confrontations, but a real analysis of needs by people equally committed to getting the best value for the Project. New estimates would be checked against the previous ones and the reason for any changes explained. These would then be collated for Wüster who would make his own assessment of how resources could be matched to requirements. Once the budget was established it was up to the Division Heads to ensure that their commitments corresponded to their predictions. Orders could not be placed without commitments being established and these would be set against the budget item. Should the money begin to run out under a particular heading in advance of completion, the responsibility lay with the Division Head, who then had to make the case to Wüster for drawing on the contingency. Given sound technical reasons, Wüster was adept at finding means of accommodating such added costs, but if the reason was not considered good enough, his strictures could be scathing and the person responsible was unlikely to repeat his mistake. Wüster's capacity for memorizing data such as are contained in budgets was phenomenal and it was idle to hope that because the budget files were not open in front of him that previous declarations could be misrepresented.

The JET Project Control Office led initially by Christopher Watson and from 1982 by Peter Kind and consisting of no more than three people (of whom at times some were part-time) contrasts markedly with the office of the same name at TFTR, which contains about 30 people. The difference lies in their basic function. In JET the office is an instrument of administration keeping the Director continually in touch with both the financial situation and the state of the programme; at Princeton it is more of a permanent watch dog put in by the Department of Energy, the main funding agency. The JET Council's control was confined to approving the programme and the related budget estimates and audited accounts. It received documents prepared by the Project Control Office (those of Kind in particular being of admirable clarity) but as formal Council papers only, presented by the Director. The sole person on the JET staff who has had any resemblance to a Council man is Jim McMahon, who from early 1979 was Secretary of the Council and the Executive Committee. Whilst in close contact with the Director, and quite clearly responsible to him, he has enjoyed a certain independence and has been in regular touch with the members of the Council and the Executive Committee. His role was to see that the minutes of the meetings were an accurate record and that at both the official and personal level, delegates felt part of the Project. His particular blend of Irish eloquence and reflective insight was an important factor in helping Wüster to build that sense of mutual confidence between the Project and its members that was essential to its well-being.

On the technical side, Rebut's control was equally personal and his detailed knowledge of all parts of the machine, as already noted, was a major element in ensuring a coherent whole. A toroidal device calls for especial care in matching systems as the smallest change in the mechanical design of one part can have a major effect on another and also on assembly procedures. In addition, the very large magnetic forces generated, notably under fault conditions, introduce design requirements that may be quite extraneous to the particular function the designer has in mind. During the design phase Rebut would supervise everything, but as new staff arrived and the detail became elaborate, a Technical Control Document was established that required the signature of the head of any section likely to be affected, before manufacture could be authorized. Subsequently there were technical control documents relating to the tender, to the contract and also to any interface necessary, the last being of particular importance in the area of diagnostics. Some newcomers saw the system as being unnecessarily bureaucratic. They had been used to smaller devices and perhaps procedures that relied principally on trial and error, and they found the discipline of the technical control documents irksome. Particularly resented was the need to re-design something because some totally different section opposed the design proposal, but everyone learned quickly that signatures could not be appended lightly with impunity, and few would not now concede that the technical control documents were a crucial instrument in ensuring that the device would go together and could be assembled.

Continual adjustment was needed to the scheduling of the technical programme and in the beginning it was proposed to put the planning on an integrated network. Culham had a computerized PERT system with bar chart display in operation and Watson sought to link this with the costing system. In practice, however, the system was found to be too cumbersome, the computer gave a lot of trouble and early in 1980 a manual system was set up in duplicate. The PERT procedure, separated from finance, was retained as a back-up, but once construction on site had begun in earnest, it was the hand drawn bar charts that constituted the planning and the PERT programme was more a public relations exercise. However necessary though such procedures are, a dominant factor in both the planning and the design was the flexibility of approach at the top and the speed with which decisions could be taken. Successful applicants for posts found it astonishing that they were offered the job at the interview even, and although it might be three months before contracts were exchanged and another three months before they arrived on site, when they did so, they found the same pace, willingness to take responsibility and, at the same time, an expectancy of performance of a high order.

Staffing was a continual problem with the result that pressure at all levels was always intense. Moreover, with Rebut demanding technical excellence, working all hours and seven days a week and continually proving himself more knowledgeable of the different techniques than the experts, and Wüster seeking always to hasten the work and contain expenditure, life was no sinecure. Nevertheless the sense of purpose and the feeling of permanent tension generated a remarkable spirit of dedication. And if Wüster's tempestuous style was bad for the nerves and the *amour-propre* at times, his essential fairness and infectious friendliness more than made up for the strain.

DEVELOPMENT PROBLEMS

Planning techniques and staff enthusiasm were, however, not sufficient to plug the two main holes in the programme, namely the development of additional heating methods and the provision of adequate diagnostics. For these JET had to depend upon the Associations as the effort involved was far beyond its own capacity. Indeed in the case of the neutral injection system, the size of the system planned and the technological problems to be overcome added up to a project of the same order as the machine itself.

Additional Heating

Neutral injection, as already indicated, is a method of heating the plasma by firing into the vacuum vessel from the outside, beams of hydrogen atoms accelerated to a high energy. It would be far simpler to produce ion beams of high energy, and indeed this is an intermediate step, but they could not be made to enter the plasma because the magnetic fields would turn them away before they ever reached it. It is necessary then to neutralize the beams before they pass into the device. Two numbers characterize the beams — the energy and the power. The energy determines how far into the plasma the beams will penetrate before being absorbed and JET determined to aim for a system which could deliver 80 keV of ordinary hydrogen and which could be extended to give 160 keV.

Although at such high energies the efficiency would be low it was believed that this was the best compromise between energy and useful power — and it is the power in the beam that determines the amount of heat put in. With such an energy the beam would be able to reach deep into the plasma even at high densities. A third number, however, determines much of the engineering design and that is the length of the pulse. JET's requirement was for ten seconds or more which meant active cooling of all surfaces where power would be deposited.

In essence, a neutral injection system consists of an ion source, followed by an acceleration stage, a neutralization stage and a divertor section for disposing safely of the unneutralised ions. In one way or another each of these three stages is incompatible with the next and the problem is further aggravated by having to compress a number of these systems into a narrow space so that finally several beams can be focussed on to the plasma through the same port. To begin with, low energy ions are produced in a plasma source that consists of an electric arc struck between an anode and a heated cathode in an atmosphere of hydrogen. Electrons dissociate the gas, forming ions which are drawn away from the plasma through a series of grids with an overall voltage drop equivalent to the energy required — or a little more as it is necessary to decelerate them a small amount in order to discourage electrons going backwards, up the beams. The grids are a compromise between providing an even potential and an adequate aperture, and being able to stand up to the battering they receive. Typical grids have several hundred apertures that need to be carefully aligned from grid to grid. The source requires an atmosphere, the acceleration stage, high vacuum. The next stage, neutralisation, is a crude process and is achieved by passing the ion beam through a gas cell where some of the ions pick up electrons (often only to lose them again soon after). As in the plasma source a variety of combinations of hydrogen nuclei and electrons are formed but this time only the fast neutral hydrogen atoms must be allowed to pass, and everything else dumped. To stop further dissociation the neutraliser cell must be followed by ducts maintained at a very high vacuum from which the unwanted and unneutralised ions of different types (carrying the majority of the power) must be turned away by magnetic fields. In an idealised system they would be turned right round and the power they carried reused; otherwise they must be absorbed in cooled traps. Neutral injection demands high voltages between closely spaced water-cooled components, high vacuum associated with gas filled cells, cryogenic pumping next to big power dumps and high thermally stressed components made and installed with watchmaker's precision — a technologist's nightmare.

During the design phase, both Fontenay and Culham were developing sources and beam lines for neutral injection, each with its own particular approach. Both laboratories were also working under contract to JET and the work was supervised by a joint committee and coordinated in JET by John Sheffield. Nevertheless, although JET drew up a general analysis of what was wanted, it could not impose a programme of work. The Laboratories were grateful for the financial contributions from JET, but were not prepared to give JET their full attention as they had commitments to their own tokamak programmes. They had not moreover the same sense of urgency as JET and the approach generally was more that of a university than of a mission oriented development.

At the end of 1977, Sheffield decided his future lay best elsewhere and he left the Project with the result that Duesing when he came from IPP to head the Plasma Systems Division during the built-up of staff in 1978, found himself in this sector, taking over a void. At the time that the grand decisions were being taken to speed up construction and double the additional heating power, JET did not have a specialist group working on neutral injection. Only in January 1979 did Ernie Thompson cross over from Culham to become the new group leader and it was not until nearly the end of the year that he had a group (of four professionals) and some order could begin to be put into the separate efforts of Fontenay and Culham. Communication was not a problem and there was a full exchange also with the Americans and the Japanese, it was a question of deciding on which of the different design concepts to choose and then to concentrate on making them practicable. It was one of the few occasions when Rebut's flair let him down. Recognising the hole that had been left by Sheffield's departure, he made his own evaluation of the state of development and in particular elected to follow the route of starting from a group of four sources each providing 120 A of ions. As time went on, however, JET was forced to accept the Associations' view that no amount of ingenuity could reduce the stored energy to an acceptable level and catastrophic breakdown would be inevitable if one persisted with this concept.

It was a source of great tension between Wüster and Rebut. Wüster had learned that the development was leading into a blind alley, but Rebut did not report the problem to him for a long time and a major split between the two men was only narrowly avoided.

Thought had been given at one time to buying sources from the USA and the Americans would have been very happy to sell as this would have allowed them to off-load some of their development costs. Their maximum energy was however only 60 keV and their pulse length was so short they could get away without water cooling. For JET, wanting 10 seconds at least, cooling was essential. The discussions went on until finally early in 1980 the broad lines of the JET design were agreed. Eight 'PINI' (at that time Plug-In Neutral Injector, but now converted to mean Positive Ion Neutral Injector) sources would be incorporated in a single injection system, and the disadvantage accepted that there would not be room to reflect the unneutralised ions back upstream of the neutraliser (so avoiding gas production in the sensitive area downstream). The Culham design of source was preferred as giving a better yield of the appropriate ions and the tasks that Fontenay and Culham would be doing, notably in the provision of test facilities, were mutually approved.

A great deal of development nevertheless remained to be done. Nobody at that time, for example, had ever produced water cooled vacuum-tight grids (and only after much research was a system based on electro-deposition found to be satisfactory). Insulation was a major headache. Plastics could not be used because of the high power involved and bonding ceramics to metal over large areas presents major difficulties. At a later stage in the manufacture every brazed joint of the copper drift tubes failed in thermal cycling tests and a totally new process had to be devised. Design of the dump magnet was very complicated because of the need to deal with different species (with a range of a factor of four in mass and negatively charged as well as positively charged ions) and of crucial importance if the final efficiency was not to be disastrously low was the cryopumping system (which 'freezes out' particles falling on it) for the ducts downstream of the neutraliser.

Nothing existed on the market that could provide the rate required, falling short by a factor of 1.5, and eventually Rebut stepped in with his own design of an open aluminium structure that proved to have a rate twice that of previous models. Not even the theory was simple. A problem arose with the ion beam focussing, for example, and this was resolved with the help of the Japanese. The Berkeley laboratory in the USA was contracted to do tests on high heat transfer elements. Indeed, the JET problems became an international concern. Yet despite the coordination and the scale of the effort, everything took more time than expected — as it did everywhere in the world where neutral injection techniques were being developed and JET was to be in operation for 2½ years before the first neutral beams could be injected.

By that time, radio frequency (RF) waves had already been heating JET for a year at a cost in development and fabrication perhaps one third that of neutral injection. Pioneers of the system in Europe were Fontenay and the first system in Europe to heat a tokamak by 'ion cyclotron resonance' (ICRH) was installed on TFR. In many respects its development was the opposite of neutral injection. Whereas the principles of the latter were rather self-evident and the first experiments gave immediate results, the science of RF heating is extremely difficult and the first experiment (on the ST in the USA) was not successful. Some idea of the problem may be grasped from appreciating that the refractive index of the plasma at the frequency of rotation of the plasma ions (the cyclotron frequency) is around 80, whilst water has a refractive index for visible light of about 1.3 and diamond about 2.7. On the other hand, whilst the technology of neutral injection at the energies, power levels and durations required is beset by intransigent problems, the difficult technological aspects of RF heating can be reduced to a few well-defined items.

Research into ICRH had begun at Fontenay before Rebut moved to Culham and he was always alive to its potential. Once the essential physics was understood and the basic principles demonstrated, JET promoted its development by placing a series of contracts with Fontenay and the results of their experiments, together with a general design for JET, were presented in June 1980. The decision to equip JET with RF heating following further development and testing was taken in November 1981. Jean Jacquinot, the bearded luminary of RF heating who, at Fontenay with Jon Adam of Belgium had mounted the first ICRH experiment in Europe on TFR, was an ardent advocate of RF heating also for JET. He came to JET in July 1982 to head the RF group and seeing this as essentially a physics problem, he assembled round him a team of physicists whose backgrounds were primarily in wave/matter interactions; few had experience of RF engineering. The outcome was spectacularly successful, programmes ran ahead of schedule, and within 2½ years he was launching megawatts of RF power into the plasma. One of the tricks that Jacquinot and Adam had invented was to tune the frequency not to that of hydrogen, which is far too absorbent, but to a harmonic of the resonance frequency of an isotope of helium which is added to the plasma and acts as an intermediary. Crucial, however, to the success was the design of the antennae which launch the radio waves across the vacuum chamber into the plasma and which must match the plasma characteristics and not introduce contamination. Two designs were prepared and both proved to be efficient under test. Before installation, moreover, complete systems were checked out, including their interface with the CODAS system so that commissioning time on JET itself was reduced to a minimum.

Diagnostics

The long over-run in time and the consequent greatly increased cost over original estimates that were experienced with neutral injection can be explained by the big jump in technology that was being attempted and the big increase in power that had been decided. So many elements in the system were at the limit of what was technically possible and so many apparently mundane components in the special environments that would obtain proved to be outside the range of current manufacturing techniques. The same cannot be said for the diagnostics the cost of which were, in R5, under-estimated by an order of magnitude.

The figure quoted then and subsequently repeated was not an idle number pulled out at random, although the tendency during the Design Phase was to minimize as far as possible all costs that were not directly linked to machine performance in order to be able to build a device with the maximum possible potential. To this extent, nobody in the Team was too anxious to question the realism of the estimates. An excuse also was that the Associations would be providing the experimental equipment and in the various meetings that had been held to discuss JET's requirements the sums had not been seriously challenged, even though some had had their doubts.

In this context it was convenient to think of the JET device as a plasma producer on which experiments would be performed whereas in all other circumstances fusion people are at pains to point out that the device is the experiment and any experimental programme is an integral part of operation. The dichotomy of view is in part explained by the optimism that was raging throughout the fusion field towards the end of the seventies. Physics experiments on JET were regarded as being of secondary importance to machine performance and the diagnostics were primarily needed to assist in tuning, in pushing up the current and the density and optimising the effects of the additional heating. Instrumentation that would measure the plasma current, volts per turn, position of the plasma and give a snapshot view of the temperature and magnetic field were taken for granted while other more complex measurements could be made when there was time to spare.

JET was primarily a device for producing near reactor conditions in a plasma, not a plasma physics experiment. Nevertheless, room was left around the machine for many types of measuring device and the Associations were encouraged to prepare themselves for an experiments programme. Prior to the foundation of the

Joint Undertaking they had the financial inducement of work of this nature for JET being eligible for priority funding and JET, through contracts, paid half the staff charges. No one however realised in the beginning how great would be the transformation from the systems they were used to, to those that would be needed for JET. Mounting a simple spectrometer operating according to well-known principles on one of the smaller tokamaks could be accomplished in a few weeks. It is perhaps exaggerating to suggest that it was sufficient to support the probe — a laser for example — and detectors in laboratory clamps, but certainly no one in this area of research had experience of the scale of engineering required for a system monitoring a device as big as JET with apparatus required to withstand high levels of radiation, largely inaccessible for days — perhaps weeks at a time and controlled remotely by computers. Apart from the scale, few had experience of preparing such systems in entirety and costing them fully. Most laboratories have large operational budgets that take care of the provision of power supplies, workshop costs, cabling and so on, and only the explicitly purchased items are noted against a particular system. Designers were therefore conditioned to counting the cost of the special probes and monitors neglecting the supporting engineering and services. All these, however, would be chargeable to JET when contracts were placed for manufacture and only then was the size of the bill appreciated.

One of Bickerton's main tasks when the Scientific Department was formed in April 1979 was to identify the different diagnostic systems that would be needed on JET and to discuss with the Associations what they were interested in contributing. By the middle of the year some 26 contracts for design studies had been placed with the Associations and as these matured a clearer definition of the possibilities became apparent. The measuring systems more closely identified with operational control, JET took care of internally. The results of the design studies were discussed in the first quarter of 1980 in a series of workshops out of which three lists were drawn up, the 'A' list contained those diagnostics considered essential for start-up operations and whose cost of 24.5 MEUA as against the 5.5 MEUA previously estimated (January 1980 prices) could be accommodated within the budget. The 'B' list covered the main systems that would be needed for plasma studies as the experimental programme developed, and the 'C' list those systems which needed further development. At the same time the laboratories responsible for each were defined and the modus operandi agreed. This provided for four phases:

Phase 1: Detailed design specification and firm cost estimates;
Phase 2: Procurement of components and materials, assembly and testing in the Associated Laboratory;
Phase 3: Installation on JET and commissioning;
Phase 4: Operation.

In this way JET was able to keep control over the development and, in particular, ensure that what was proposed was indeed compatible with the machine design—and other diagnostics, before parts were ordered, and that performance did match the specification before installation began. However, no costs of Phase 1 fell on the JET budget, these being covered by the Associations and Euratom and the result was that less than top priority was given to the JET needs in a number of cases. Once it came to Phase 2, JET was paying for full costs and could, in consequence, exercise a firmer control.

During 1981, Phase 1 was completed for most the 'A' list and several of the diagnostics were well into Phase 2. It was already evident, nevertheless, that it would be difficult to have them all ready for start up in mid-1983. In mid-1982 the extra funding needed for the 'B' list was approved and soon after the Scientific Department was reorganised to match the situation. Until then it had consisted of one division only (Experimental Systems) led by Gibson, divided into three groups of which one only, led by Peter Stott, latterly of Culham, was responsible for diagnostics. Under the new organisation, Gibson became Deputy Head of Department and three Divisions were formed of which two were to look after different types of diagnostic equipment, one under Stott and the other under a new arrival from IPP, Wolfgang Engelhardt. The third Division, also under an IPP man, Diethelm Düchs was the Theory Division, and it was formed despite some opposition from the Associations which saw no need for theoreticians to be established within the Joint Undertaking as, it was argued, they could just as easily work in their home base. Experience showed otherwise. In the first place, distant researchers tended to work on their own problems independently of JET and with little sense of time. In the second, interpretation of measurements was a major activity essential for the day to day operation of the machine.

Few of the diagnostics give simple direct readings of a parameter independent of other parameters. The majority produce numbers such as the frequency and intensity of a spectral line and to convert these into definite plasma characteristics such as density or temperature at a particular place and at a particular time requires a great deal of theoretical understanding of the conditions under which the measurements are made. Diagnostics were designed around interpretative models that had been elaborated on other fusion devices and for the most part they were built to be technician operated. This meant that nominally they would give values of plasma parameters—provided the general conditions were those that had been assumed in the model. Should these be different, especially if in an unknown way, then the theorists had to be called in to analyse the problem and see what really was happening.

CONSTRUCTION AND COMMISSIONING

With the persistent shortage of staff, the compressed time scale and so the constant pressure on everybody, the technological problems to be solved and the inherent complexity of putting together a massive and hugely complicated ball (more or less) with so many components embedded in or lost to sight under others, final manufacture and erection seemed a recipe for continuous chaos. In the words of one of the senior members of the assembly staff 'When I joined the Project early in 1979 and realised how much there was to be done, I was convinced it was impossible and we should be doing very well if in the end we were only very late. Yet when it came to the point, things fitted just all the time and in time'. Indeed the erection phase was characterised mainly by the absence of catastrophes and the quite extraordinary precision with which the main machine components came together.

This was the result of neither luck nor accident and the general opinion on why the machine construction was such a success can be summarised as follows:

1. The Project had absolutely clear direction. Rebut was familiar with the details of the whole machine and would give immediate and definite decisions when any design point was in question. Wüster would do the same for organisational problems.
2. The main machine components had been given a great deal of attention and in the waiting years, design had been continued. Again to quote a member of the assembly team 'only where the design team had been uninterested or there had not been effort available to put in the detailed thinking, did things have to be cobbled together. Where components had been considered in depth, the engineering result was impeccable'.
3. The same people were responsible for design, establishing manufacturing contracts, supervising manufacturing standards, installation through to final testing.
4. The system of control documents, although cumbersome and frustrating, guarded against incompatibilities.

Not that the construction period was without incident. The pace was grinding and at times the strain on particular members of the staff approached the limit.

Reflecting on the problems posed by cabling and pipework, one senior planner recalls — 'these were the worst six months of my life; I hope I never have to go through that again'. Nevertheless despite the intensity of the demand made on people, and no doubt in part because of it, the spirit in the Project was one of total personal dedication. For many, the whole scale of the enterprise was a new experience and, in consequence, it was the challenge that took hold of them, but those who came in from outside and were familiar with big projects such as nuclear power stations, became equally caught up in the drive and were carried along by the urgency that was all around. Their knowledge was of great value to the Project and during the construction phase their standing was high. Later they were to be reminded that they were, however, not really fusion people and like so many mercenaries before them, found their status rather diminished once the objective had been achieved — when the device became operational and the 'real' fusion people had developed their own confidence.

1979

Site work began in March 1979, the same month that 'planning permission' for the work to proceed was received from the competent local authorities. Winter had been severe and the bad weather continued well into the Spring, so JET got off to a slow start. Complicating the initial civil works was the high water table and low load carrying capacity of the ground, and amongst the first tasks was the enclosure of the excavation area in an underground wall and the drilling and casting of piles. Priority for buildings was given to the Control Building and the Assembly Hall, and on 18 May Brunner laid the foundation stone of the JET specific buildings. This is in the wall of the Control Building (in different sources referred to as the Main Building or Main Experimental Building). The date had been chosen to coincide with the Culham open days when a great number of visitors as well as the families of staff came to see round the establishment and hear about the work going on. The weather improved, site work accelerated and in August the main contractors were able to take possession. They were immediately thwarted in their plans by a strike in the steel industry which held up delivery of steel frame sections. As a result only the Control Building could be completed within the year as planned. However, the work was rephased in part by re-ordering the construction schedule and in part by moving on to other buildings. At the same time the 'non-specific' JET buildings to house the staff, which were the responsibility of the Host, took shape and were all roofed by the time the hard weather set in.

On the machine side, while the detailed design of ancillary equipment continued and the pace of tendering and placing contracts built up, prototype work for the main components was vigorously pursued and first production runs began. Many physicists to their surprise found themselves spending more time in manufacturers' works dealing with engineering questions than at JET. Although a quality control group was set up, its numbers were kept deliberately small (4 people). Its functions were first to advise and help the contract liaison officers in the engineering divisions to prepare specifications, choose standards and manage the quality control aspects of the manufacturing contracts. In addition it monitored the general level of control achieved, prepared standard specifications for specialist materials or components (particularly valuable in relation to vacuum techniques) and established internal procedures and documentation. Four levels of control were identified, the most stringent relating to components whose failure would cause a severe interruption to the operational programme and the lowest relating to commercial supplies where it was only necessary to confirm that items delivered conformed with the makers' specifications. In effect, the group provided the experience and it was then up to the Divisions to take advantage of this to ensure that what was delivered was to the standard required.

A major event in November 1979 was the delivery of the first four toroidal coils—the first solid proof that it really was a tokamak that JET was all about, not simply a mass of drawings and a collection of buildings. The event like many others to follow was feted appropriately, continuing a tradition established by the Design Team whose Christmas parties, for example, had become renowned. Wüster lost no opportunity to mark the progress that was being made and to emphasise the team concept. The arrival of main components provided an excellent excuse and not even the absence of the principals could dampen the enthusiasm. The delivery of the first vacuum vessel octant was celebrated in champagne, but not by many of the group concerned—they were already on their way back to the manufacturers to keep up the pressure.

At the beginning of 1979 the computers to be used in the Control and Data Acquisition System were selected and orders placed. Originally the company chosen, being centred in Norway, had not even been circulated, but their contention that they were manufacturing in France was accepted as sufficient qualification for them to be considered. Being of the same mark as those adopted for the SPS at CERN, collaboration with the relevant Division there was particularly valuable even though the similarity between the two control systems is rather superficial. JET was able to make a start with system testing and development using CERN's interpreter and other software facilities and their general experience and expertise were much appreciated.

With the broad tokamak requirements by then defined and the work becoming more specifically control oriented, Franco Bombi took over from Noll as Head of the CODAS Division. Bombi's own field was automatic control and he had come to know JET through a contract the Project had placed with the CNR late in 1975. This had brought him into contact with the staff with whom he remained in touch. He had joined JET at the beginning of 1978, becoming full-time in the Summer of that year.

A major task over this early period was to coordinate all the activities of the other Divisions concerned with control instrumentation and data acquisition, and to establish specifications for the common and standard units that would be employed throughout the system. After that tenders had to be sent out and orders placed.

In the Summer of 1979 a study contract was placed to define construction procedures for putting the JET device together. In broad outline these had already been established, but it was still necessary to confirm that there were no inherent snags and to write down each step in the complex series of manoevres that the curious configuration of JET imposed. It was also necessary to design and put out to manufacture all the special tools and jigs that would be required for the assembly. Still to be defined even in a number of instances was the extent to which assembly of a component should be performed at the manufacturers or on site.

Had JET not been designed to be maintained and modified by remote handling techniques the machine would have been so much simpler, and the tolerances on components could have been much relaxed. It is difficult enough to build a nest with components lying inside others and all bent round into circles, with individual units weighing 60 tons and more; it is very much more difficult to arrange at the same time for any part to be replaceable using remotely controlled equipment. This implies that the major elements must conform exactly to a specification and not be individually made to fit.

Over and above the purely technical aspects of the assembly, the problems of manning had to be considered. Whereas JET people could be expected to work all the hours there were at whatever most urgently had to be done, contractors' staff would be subject to different trade union practices and safety conventions and would also want to have civilised working hours. Not many of the JET staff had direct experience of such problems, having been brought up in the more flexible atmosphere of a research centre.

The report that emerged from the study brought to light no insuperable problems and it became the base document on which the Main Assembly Contract was formulated and let.

1980

During 1980, manufacture of the toroidal coils settled into a steady rhythm and the crucial interface material between the vertical limb of the D and the inner cylinder was selected. The prototype poloidal coil for the machine axis successfully passed all its tests and by the end of the year five out of the eight had been wound and impregnation of the first production coil completed. Winding of the first half coils Nos. 3 and 4 was done to specification, but problems were encountered with the moulds for impregnating them and their expected date of delivery moved back to April 1982. The lower coils had to be installed in temporary positions early in the assembly programme before any of the vessel and toroidal coils were moved in (to be lifted up round the torus after this is complete) and so the delay was disturbing.

Production was going smoothly on the laminated transformer limbs and good progress had been made with the castings for the mechanical structure. Some defects had been encountered in the first castings and production was held up for a time while the cause was analysed. This resulted in major improvements being made to the casting procedures. Later porosity outside the specifications was found in the shell castings after several had been made but happily tests proved that the specification had been unnecessarily severe in this regard and the castings could be accepted without risk.

There was encouraging news also of the vacuum vessel following the disturbingly slow start, although further delays were threatened as a result of the Naples earthquake of November 1980 which had damaged the building in which part of the bellows contract was being carried out. Nevertheless, during the year problems of manufacturing tolerances for the bellows had been overcome and several units had been delivered to the manufacturer of the rigid sections who was also responsible for the sub-assembly into octants. The decision had been taken to subject all octants following delivery to JET to a bake out to 600°C under vacuum in order to reveal any leaks or release stresses that might otherwise only become evident when operation began. To this end an oven of internal dimensions $6.1\,m \times 8\,m \times 3.9\,m$ had been ordered and the parts for this were delivered in September.

Contracts were placed for the majority of the remaining electrical sub-systems and towards the end of the year major installation work on site began, notably of the two flywheel-generator-convertor units and the 132 kV/11 kV sub-station. The 400 kV sub-station for which the Authority and the CEGB were responsible was also in an advanced state.

During 1980 also the control building was finished, the majority of the computers were delivered and tested, and the terminals and screens in action gave the impression that commissioning had already begun. The Assembly Hall was clad and roofed and installation of equipment inside could start, although the Hall was not yet ready for use. The real mile-stone of the year was the handing over on 16 October of the offices and small laboratories that would house a large number of the JET staff. Not all, the temporary buildings which had been their home until then would still be required as the staff numbers built up. The formal hand-over from Sir John Hill, Chairman of the Authority, to Professor Teillac, the Chairman of the JET Council, was a joyous occasion marked by a ceremony at which an appropriate plaque was unveiled.

This was the moment when the Joint Undertaking could really begin to sense its independent identity and people could feel they were part of a common single enterprise separate from the host laboratory. Wüster no longer had to grind his teeth in silence when he found Pease conducting visitors round his buildings, and he made it known in clear terms that from then on, this was his organisation, his site and anything to do with it went through him. His presence was progressively more strongly felt within the JET staff. Communication became much easier and with his phenomenal memory for people's names and their individual problems he was rapidly known to and by everyone. The upper front corridor of Kl (the main office building) echoed to his booming voice and his ready laughter and also his ferocious outbursts. He involved himself much more in the technical side, attending weekly planning meetings and taking a firm grip on technical policy.

For the machine itself and all the auxiliaries, responsibilities were even more widely distributed as Division Leaders, Group Leaders and Section Leaders took possession of their own hardware and continued with the development of methods and components still to be defined. Rebut was everywhere, involving himself in the detail as well as the grand design, as concerned with the physics implications as with the engineering. His overall grasp of the science earned him the Grand Prix de Physique Jean Ricard which was presented to him by the Société Française de Physique in June 1981.

1981

Throughout 1981, more and more components, including all the outstanding toroidal coils, No. 2 lower poloidal coil, the lower ring and collar and first octant shell were delivered to site and storage problems became acute. Every spare space in Culham was requisitioned but even then not everything could be kept under cover. Wüster would not however sanction hiring any temporary accommodation. The transformer limbs had been held by the manufacturers as long as they could, but finally they were compelled to make delivery. No storage room was available for such massive items at Culham and they passed the early Winter out of doors, incidentally posing a nice problem for the handlers. No cranes capable of lifting 80 tons are available out in the open and the beams had to be unloaded from the transporting trailers, and subsequently reloaded, using jacks.

To aggravate the problem of the buildings running behind schedule, under pressure from the AEA concerned about the long-term problem of dismantling, a late decision had been made to change the design of the roof of the Torus Hall and the adjacent radioactive test area. The roof instead of being poured as a continuous slab was being constructed from beams cast *in situ*. This meant that all the area underneath remained encumbered over many months with massive scaffolding that blocked off any passage. In order to circumvent this problem, part of the wall of the Torus Hall was left open, so goods could be moved in through the gap. One should note that not everything connected with the buildings was behind time nevertheless, and the two massive shielding beams each weighing 1200 tons and the 350 ton shielding doors separating the Torus Hall from the Test Area and the Test Area from the Assembly Hall were completed on time.

Inside the Assembly Hall, the jigs necessary for the mounting of a complete octant had been assembled and special tools had been tested. The main 150 ton crane, although late and far from trouble free, was commissioned just in time to unload the first shell octant which was delivered to JET in November. Over the next few weeks, additional work was carried out on this including the application of instrumentation sensors and by the end of the year it was ready for the toroidal coils to be inserted. In a nearby area the vacuum oven had been commissioned and tests carried out first on the prototype octant vacuum vessel which was delivered to site in July, followed by No. 1 of the production series. These showed how the testing technique could be improved and also how necessary such tests were because they revealed leaks between the inner skin and the main volume which had not appeared during partial room temperature tests at the manufacturers. The experience gained in preparing the prototype had resulted in major improvements being made to the quality of the bellows assemblies, now rolling off the production line, and octant assembly had also been stream-lined. As a result, there was growing confidence that despite each octant requiring 1 km of weld in an inherently difficult material, the extremely stringent vacuum requirements would indeed be met.

Installation of the big electrical components was proceeding well and the different power supply buildings were looking more and more like the inside of a major power station. The 400 kV sub-station was completed and the two main step-down transformers, having first been tested at the manufacturers, were on site. Assembly of the flywheel-generation-convertors was also gathering momentum.

Access to the Torus Hall was finally obtained in October 1981 and to the sound of bulldozers still working next door, the system of 32 datum points was established (at 8.5 m and 13.5 m from the machine centre). These were the reference points used to position the main components with high precision. Here again the toroidal shape introduced peculiar requirements, and normal surveying or line of sight procedures had to be adapted to the three dimensional geometry. JET was calling for a vertical alignment accuracy of 0.2 mm and the angular radial tolerances demanded of the datum points was 0.25 mm. Subsequently the torus hall was seen to be sinking and tilting and the whole machine dropped by about 5 cm, while the machine vertical moved by many millimetres. The relative positions of the machine components were not, however, affected, and the surrounding floor was built up to align with the floor of the Assembly Hall.

In November, laying of the machine base plate began and one could say that erection had really started. The occasion was marked by two 'Families Days' during which JET and the Culham Laboratory entertained some 2500 visitors who were able to tour the almost completed buildings, see the Assembly Hall, by then crammed with equipment, and even go into the Torus Hall through the diagnostics aperture in the south wall.

1982

The year 1982 was one of intense excitement. This was the time when it would be known whether the fine tolerances that had been demanded really had been achieved and whether all the multifarious components really did match. Conditions were still not ideal; only temporary heating was available in the Assembly Hall and, in the Torus Hall when the building was handed over on 15 January and they began to install the large cooling manifolds under the machine, followed by the bottom transformer limbs early in February, the temperature was $-6°C$ inside and $-10°C$ outside. The latter task was not eased by the failure of the turning motor controlling the crane hook which consequently had to be disconnected, but within 10 days all eight were in place. The low temperature also created problems when installing the base plate for the inner poloidal field coils as this was grouted into position with an epoxy resin that refused to set. By the middle of April nevertheless all the vertical limbs had been set up and the bottom centre piece was in position and the time had come to build the machine itself.

Out in the Assembly Hall, the first octant had been slowly and carefully put together, a process that started with half an octant shell being laid on its back in a cradle, i.e. with the open side upwards. Two toroidal coils could then be lowered into the channels provided and there fixed in place. Returned to the vertical position the sixteenth was mounted on one arm of a turn-table and subsequently a second completed sixteenth was mounted on the other. In between was one octant vacuum vessel already lagged and instrumented and round this the two sixteenths were closed and bolted together. As a plan, quite elementary, but it took three days simply to turn the first sixteenth through the 90° necessary. Altogether it took seven months to build the first octant—and six weeks to build the last.

To prepare an octant for insertion in the machine, numerous small jobs were still required and for these to be done the octant was removed from the assembly jig using rigid stiffening frames bolted on each side of the mechanical structure and transferred to another station. There electrical connections and water manifolds were fitted and tested, penetrations for diagnostic probes, limiter guide tubes and gas feed pipes were welded on—the welds all being made inside the vacuum vessel.

Some instrumentation however could be installed only when the torus was complete. A notable example is the series of wire loops that are used to measure the voltage applied to the plasma and which encircle the torus at different heights. These are inside the shell and required a plastic draw wire being pushed round the major circumference through the matching guide tubes already in place in the eight octants, a distance of nearly 30 m. In only one case out of eight did it prove impossible to complete the circuit.

In the Torus Hall, at the end of April the lower poloidal coil No. 2—the most hidden—was placed round the centre piece and in May, the lower ring and collar, pre-assembled in the Assembly Hall to save time was transferred bodily to the Torus Hall, and mounted on the transformer limbs. The next important step was the mounting and accurate positioning of the upper ring for which four massive columns were erected symmetrically disposed around the machine centre line. The two rings, in effect, defined the geometry of the whole machine and the columns were there temporarily to lock the structure in place while the first four octants were being installed. These then would take over the task and the columns could be removed to allow the remaining four octants to be brought in.

The inner cylinder was assembled segment by segment and towards the end of July, the lower coils 3 and 4 although still unfinished were dropped down on to temporary supports on the lower transformer limbs.

Because of limitations of space at the contractor's Augsburg factory where these coils were being made, the lower No. 4 had been shipped to Mannheim for pre-assembly, so beginning a transport saga that had the JET team responsible biting its nails with anxiety. Eighty tons in weight and 11 m in diameter, the coil after checking was split again into two halves which were then loaded on to a barge for Rotterdam. At Rotterdam they were transferred to a RORO vehicle which made the crossing to Felixstowe. From Felixstowe the convoy started on its way to Culham seeming intent on making a tour of England in the process. A direct passage was impossible and it first went to Birmingham, followed the motorway down towards Bristol and then came east again on the M4 finally to join the Newbury-Oxford road and the connection to Culham. Replacement of the cooling manifolds was then found to be necessary and the upper coils were shipped to JET without pre-assembly so that the intermediate stages could be more closely controlled.

The great day when the first octant was inched through the dividing walls, its 120 tons resting in a C-shaped frame which maintained it exactly vertical (the side frames had been removed) was 4 August 1982, holiday time. Rebut was on his boat unmindful even of the article on Wüster which had appeared in the *New Scientist* on 20 May and which had so upset him as it referred to Wüster as 'the man who built JET'. (The *New Scientist* made ample amends in its issue of 23 May 1983 with an article on Rebut entitled 'A fusion of talents'). Huguet had come back to the site especially, to the chagrin of Salpietro whose particular responsibility it was. He need not have bothered. Slowly the giant load was slid into position between the vertical transformer limbs and nested into the upper and lower collars. As the whole assembly finally came to rest, the waiting fitters leaned over, inserted the holding bolts and were able to do them up by hand, such was the excellence of the mating. Everyone present, contractors' men included, burst into applause. As one of them remarked: 'never in the whole of my experience have I seen its equal', and he was not a young man. The celebration that followed was appropriately animated.

With increasing experience, the rate at which octants were assembled and finished built up, and it was perhaps not surprising that there should be one accident on the way. A lifting strop became pinched without it being noticed, broke when the load came on and a vacuum octant assembly clattered to the ground. This could have been catastrophic for the programme had it not been possible to reschedule the installation order and make use of the prototype. In the meantime, the affected octant was returned to the manufacturers, (who were due to complete all deliveries in July) where the damage was found to be relatively minor. A joint rigid sector and a bellows assembly were replaced and the unit returned to JET in October.

By the end of that month the fifth octant had been installed, closing the gap between two already in place for the first time. Insertion was eased by evacuating the interspace between the vessel walls so causing the bellows to shrink and marginally widening the space available. Such was the confidence by then, Wüster had invited Teillac, the President of Council, and the Lord Lieutenant of Oxfordshire, the leading representative of the region, to be present. Again the fit was superb, and despite protests from his staff urging him to wait, Rebut gave the order for welding up to begin at once. On 13 December came the moment of truth; the eighth octant was eased into place. Nothing to add or subtract on either side, the tolerances were comfortably within specification. The torus was closed.

Elsewhere on the site, the numerous components of the electrical supplies were being assembled, the most spectacular being the static conversion system fed by the 400 kV/33 kV transformers, each capable of delivering a pulse of 300 MVA, and the two identical flywheel generators each capable of a peak power output of 400 MVA and a total energy per pulse of 2600 MJ (roughly 700 kWh in 20 seconds). JET was installing the largest AC/DC conversion system in the world. JET's power requirements had been steadily rising and with the enhancements envisaged would be making demands that would soon tax the initial installation to the limit. The size of the load can be gauged from the fact that JET is the only customer dealing directly with the UK Central Electricity Generating Board. All others are supplied through the different regional authorities. On their side the CEGB were well on schedule bringing power to the site. The 132/11 kV sub-station had been in operation from the beginning of the year and the 400 kV sub-station with its main circuit breakers became operational in July.

Building and commissioning the generators was not without incident. They were an extrapolation of previous experience and although extensive computer analyses had been made of the stresses to be expected and the fatigue problems to be overcome, with such a massive unit it was difficult to anticipate all eventualities. Basically each generator consists of a vertical shaft, driven at the top by an 8.8 MW pony motor, on which is mounted through the intermediary of a 12-armed spider an annular flywheel outside which are the rotor poles and windings of the generator. Surrounding these is the stator with 48 poles in which an AC current is induced of maximum frequency 90 Hz. The dimensions are impressive. The rotor is nearly 10 m in diameter and the rim, which is laminated and built up from plates with axial bolts clamping it together, is over 2.5 m deep. All told the rotor system weighs 774 tons. Maximum speed is 225 rev/min which will reduce to half during a full current pulse. The pony motor takes 9 minutes to wind the rotor back up to speed. Maximum lifting capacity available in the generator halls was 50 tons and so it was necessary to build the generator in situ. One of the first things to happen was that one of the shafts weighing 40 tons was dropped and had to be returned to the manufacturers for repairs and checking. This was particularly disturbing as, in effect, the generator is built up round the shaft once this has been lowered on to the pad bearings that support the full rotor weight.

Actual erection began in February 1981 and by the end of that year the rotor of the poloidal field generator had been completed and both stators were ready. However, in the light of experience with a similar generator elsewhere, the manufacturers took the decision to replace all the stator bars and rewind the stators. These were ready in the first half of 1982 and over the rest of the year the full assembly was completed and the various auxiliaries such as the excitation system were commissioned. Commissioning of the complete poloidal field generator began in November and on 17 December the generator was run up to 20 rev/min for the first time.

Although the spider carrying the flywheel itself weighs 50 tons, it is a relatively light structure in comparison with its whirling load which expands outwards around 2mm when turning at speed. Balancing the system is thus of critical importance and calculations had indicated that resonance would set in below 300 rev/min. Traditional methods of fitting balance weights to the spider arms achieved a low level of vibration on running up to modest speeds, but suddenly at about 180 rev/min, violent vibration set in, the level of oil in the guide bearing dropped and the manufacturer's representative sprang for the brake - only to be restrained by Michel Huart, the JET man in charge. In the words of Bertolini 'there then followed the longest hour of my life'. A million pounds were at stake and at least six months in time. The arguments were nicely weighed: apply the brake knowing that further imbalance would be introduced but hoping that it would be sufficiently short-lived for the equipment to hold up, or allow the speed to drop naturally and let the vibrations die, trusting that no permanent damage would be caused in the interval. Happily none was. The reason for the vibration was identified as magnetic imbalance in the pony motor which was the largest asynchronous motor that the manufacturers had ever built. Although the rotor was balanced mechanically, the magnetic forces were asymmetric and at a certain speed resonances developed. It was Rebut who offered the solution. Rejecting the complex mechanical modifications that were being proposed he suggested instead that the stator windings of the motor should be cross-coupled in a diametric pattern. The effect was dramatic and when final balancing was completed by adding small weights to the rim of the rotor rather than the spider, the generators ran considerably smoother than had ever been foreseen.

Troubles were also experienced with the main pad bearing: the temperature rise in the oil lubrication was outside the specification and the fear was that the overheating would give rise to dry spots, further heating and a runaway situation ending in catastrophic failure. At least it was accessible, as jacks located at the bottom of each generator pit could take the rotor load and lift it clear. Initially JET was obliged to restrict the maximum speed to a level where it was considered safe to run. The device could be powered — quite adequately for commissioning and initial operation, but full performance would not be obtainable.

Again it was JET that provided the answer, this time Huart who proposed that cool oil be injected into each pad instead of relying on bulk mixing and when subsequently during a shut-down of the machine, this modification was made, the generators really could do all that was asked of them.

1983

In the months leading up to the target date of June 1983, activity mounted to fever pitch. At every turning it seemed one would find local computers linked to mobile control desks as different parts of the system were checked out. It would take far too long even to list the sub-systems involved, many of which were major engineering installations in their own right. The vacuum system, for example, with turbo-molecular pumps coupled to the torus by two large pumping chambers and backed by a roomful of roughing and intermediate pumps; the torus heating system requiring a major gas heating plant; all the different cooling systems, finally to dissipate all the power that was being fed in ... Major assembly work was also going on for the neutral injection testing plant and the various components of the injection system, all with the same urgency.

Cabling was a nightmare. Little effort had been available to design it properly, the contractor was overtaken by the scale and speed of the demand with the result that cables were being pulled in advance of their exact destination being known and some muddle was inevitable. So strongly indeed did the final tangle offend Rebut's sense of order and quality that at a subsequent shutdown he had the whole of the basement below the machine cleared and new cables run.

Following the placing of the last octant, the upper outer poloidal coils were posed and already in January work began on closing the magnetic circuit by fitting the upper transformer limbs. Inside the torus, the welding of the joints between the octants was systematically pursued. When these had been completed, the vessel was cleaned manually by high pressure jets of water and warm detergent and rinsed in demineralised water. Only then were the bellows protection plates fitted and a final rinse made using an automatic sprinkler system.

By the end of May, the torus was ready to be pumped down and for the first time its 8 km of weld would be tested all at the same time. Two leaks were discovered in external welds that were quickly repaired and straight away the pressure sank to a low level, limited only by the residual occluded water. Heating of the vessel to a temperature of $120°C$ drove off the majority and improved the vacuum by a factor of six. The dominant residue was then hydrogen.

A major scare at this time was the discovery of a short in the toroidal coil system, something potentially catastrophic as just one faulty coil would mean breaking open the torus and extracting and dismantling the offending octant. Huguet and his colleagues worked through the night trying to localize the fault, finally to stumble on the cause as lying in the coolant. The water had been sitting in the channels for some time and had become stagnant and conducting. Starting up the circulation system was all that was needed to cure the problem.

Much had still to be finished, the heating mantles on the torus ports had not been coupled up, even the gas heating plant was running on air instead of carbon dioxide and with a temporary blower unit, a minimum of diagnostics was operational, but by the third week of June, all the essentials for a first try at a plasma were functioning. First however, further cleaning of the vessel by a glow discharge with the walls at 200°C. Electrodes had been provided in the vessel for this and over 30 hours the torus was filled with fluorescent light that licked the walls and drew off more of the noxious contaminants.

By Saturday 18 June commissioning had progressed to the point that a sufficient number of sub-systems had been coupled together through CODAS for trials to begin on the device as such. The Torus Hall was evacuated, the big doors closed and the run up to discharge could start. Like launching a space vehicle, there is a long count-down before each discharge and when faults are found, they must be repaired and the whole procedure started over again. During all that week, while commissioning of further sub-systems continued, the 'bugs' were located and corrected — nothing dramatic, often simple things but time consuming and frustrating.

Come Saturday, June 25 and it looked as if they really could try for a plasma. Not quite the scenario to be expected from science films where the senior staff is grouped round mysterious screens as a disembodied voice solemnly counts down from ten. In the control room itself, there were only a few people sitting at the consoles operating the machine controls. Power supplies were being looked after from a local console far away, while Wüster and Rebut were in the Diagnostics Hall (next to the Torus Hall) where temporarily a few desks with television screens had been pushed together. Communication was assured by walkie talkies. Not a great crowd had assembled; most people were there because they might be needed. Huguet and Green sat before the screens, the others moved around chatting and waiting. Apart from Wüster and Rebut, there was Daniel Cacaut from Magnet Systems, Ber de Kock, Leader of the Plasma Boundary Group coming in and out, there was Jungen Dietz, leader of the Gas Handling and Wall Technology Group, Brian Ingram, Leader of the Installation Group, Last, Leader of the Poloidal Field and Instrumentation Group and Vallone who had dropped in to see how things were going — about 20 people in all.

More frustration, as further problems were sorted out, more hanging about. The detection systems in place were primitive. A camera was perched on top of the device, looking down into the middle of the poloidal field coils, for the time being minus their central iron core, just to check that nothing untoward happened; one screen was connected to a telescope looking through a port into the plasma, but the centre of attention was a simple Avometer propped up on the top of a cabinet. This was connected to a search coil, inside the device and should give the first indication of a plasma. Shot number 73 and the count-down began again, still going, no problems—all the way—at last—poloidal field coils energised. The centre image distorts as the camera strains in the field, but the upturned faces looking at the AVO turn away in disappointment—had it kicked?— perhaps but backwards. But there on the recording oscilloscope was the signal of a light pulse. Smiles everywhere. Phil Morgan's telescope had seen a plasma. Polaroid pictures were taken and admired while everyone waited for the results from de Kock's other pickup system. Four long minutes and in he comes from the control room waving a plot of the applied voltage and the detected current—a mere bump on a curve, but the bump was the plasma current on top of the current induced in the vessel wall. Not very big, about 16,000 A for $1/10$ second but not bad at the first real attempt.

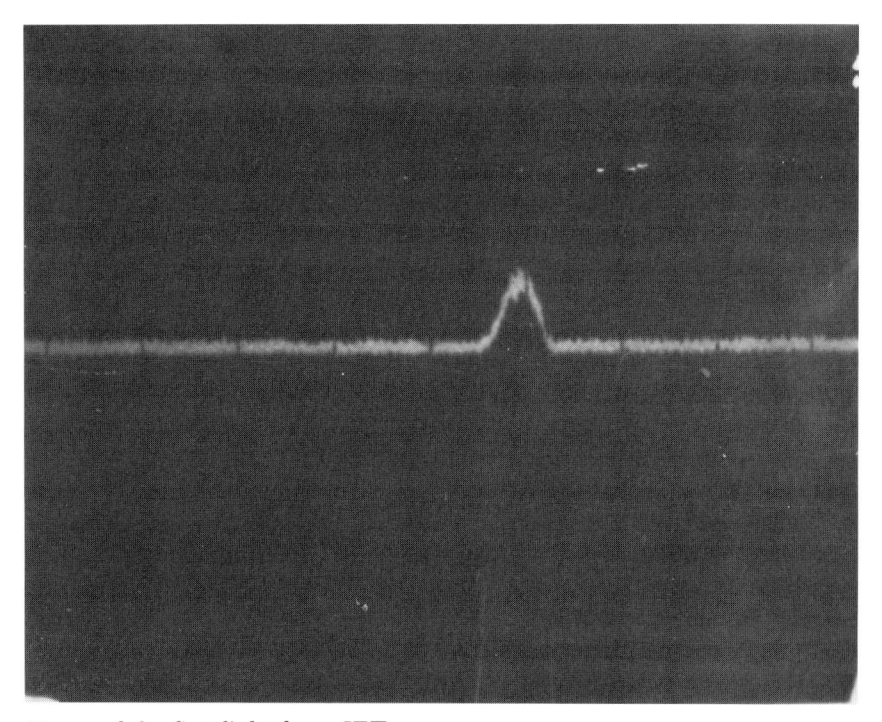

Dawn of the first light from JET.

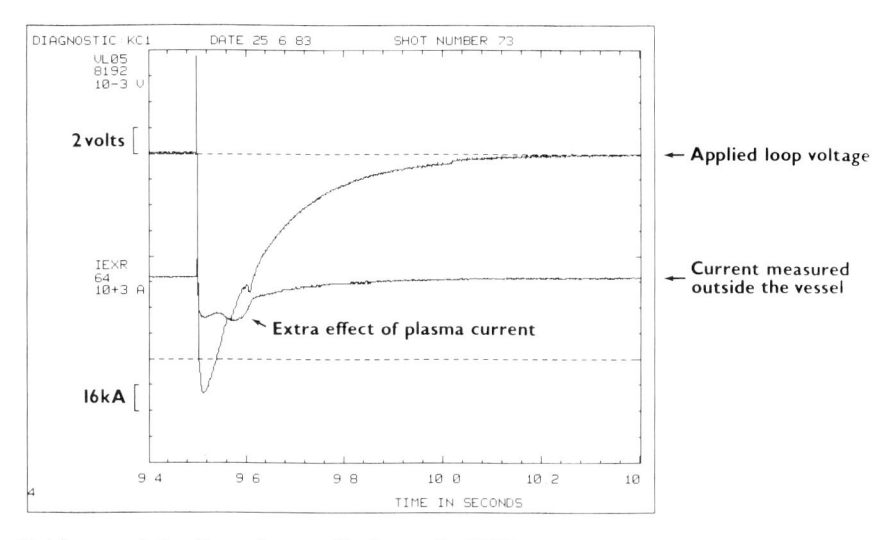

Evidence of the first plasma discharge in JET.

Try again—nothing, was it just a fluke?—and again—a pulse and bigger—and again—going up. By the end of the day, four successful discharges out of seven had been recorded and on the Sunday it was 13 out of 15, and a current of 60,000 A had been achieved. This was the number used in the low key press release that went out after Council members had been telexed the news. In fact the figure was nearer 100,000 A as the instrument was shown to be reading low by a factor of 1.7. So the first discharge was really nearer 27,000 A and contrary to the belief of the cynics watching, the Avometer had neither lied nor been connected the wrong way. It was simply that the pick-up from the poloidal field that could not be estimated in advance, proved to be in the opposite sense and bigger, and so had swamped the plasma signal.

The timing was perfect. The Council was due to meet at the end of the month and on 30 June a grand barbecue had been planned for the delegates and all the staff. It was a huge success and warmly appreciated—not least by the contractors' men on site who had never in their past experience been thus recognised by the client of the moment. Nor were the Culham people providing services to JET forgotten. Culham's kitchens organised the two roasting pigs and the accompanying dishes (and were honoured to do so), Germany sent a gift of beer and Ireland, Guinness to supplement the local brew. The weather was kind enough—fine even if typically somewhat windy—and the tug-o-war teams strained in amiable disarray.

In the one serious moment, Teillac as President of the Council referred to the press release which 'spoke of modest success—but, 'he went on, 'this was not modest achievement'. He expressed the thanks of all the Partners to the JET staff for their devoted work and to Culham for having provided 'unstinting help'. He continued by saying that 'while we did not admit to there being any competition between the Americans, the Japanese and JET, it was nevertheless very satisfactory to note that we are now only a few months behind TFTR when at one time it was a year'. TFTR sent a cable of congratulations adding they had now reached 1MA with their machine — a last defiant gesture as they knew that soon the numbers coming out of JET would be the world records.

Throughout the celebrations Wüster took every opportunity to pay tribute to 'the international team assembled at Culham, to the work force which constructed the site and assembled the machine and to the numerous branches of European industry for their enthusiastic cooperation and expertise', and above all to Rebut 'who led the design team and was responsible for the construction of the machine'. The letter he sent to Teillac in November, by which time JET had produced plasma currents of over 1.0MA lasting for over a second, placed on record 'la part capitale prise par Paul Henri Rebut dans ce succès'. He dwelt on his outstanding intellectual capacities as both physicist and engineer and his remarkable qualities of leadership, and attributed to him the credit for the discipline that had been exercised throughout the construction phase and had resulted in the machine being built without one serious accident to mar the adventure.

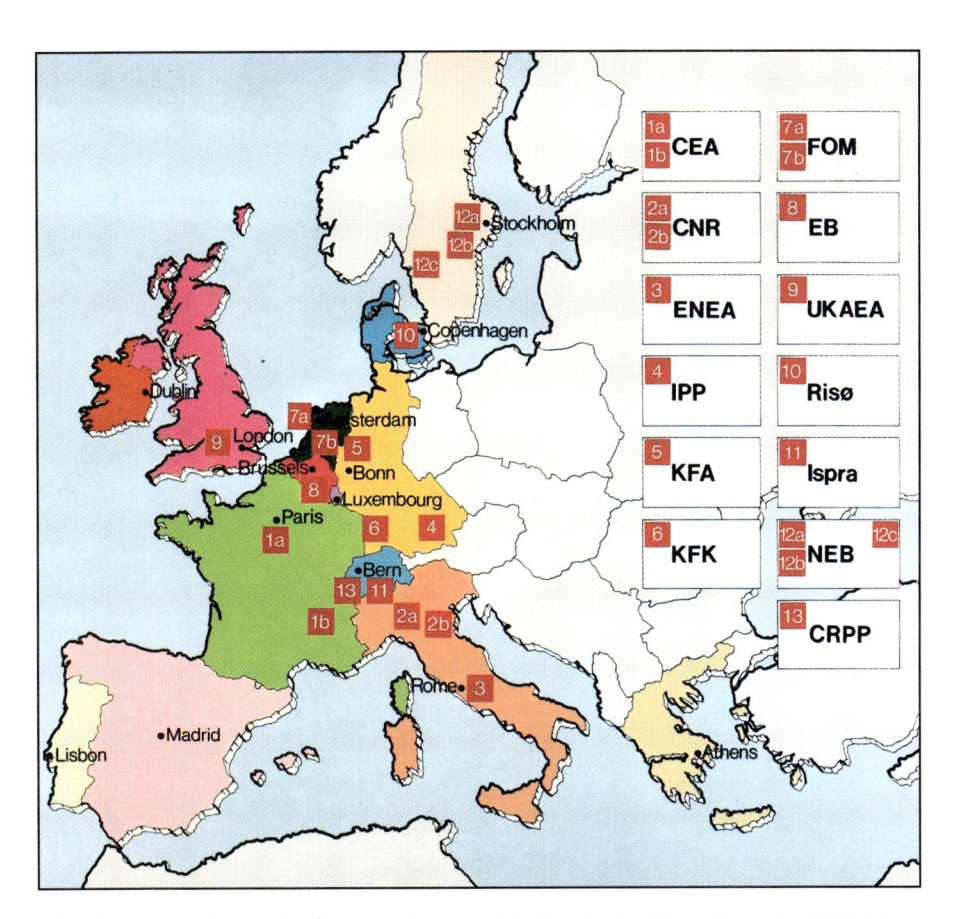

The Partners, who made the experiment of fusing their efforts into the JET Joint Undertaking, were governments and research organizations in the European Communities and Euratom, one of the three original Communities set up under the Treaty of Rome in March 1957. They were joined by the Swedish National Board for Energy Source Development and Switzerland represented by CRPP.

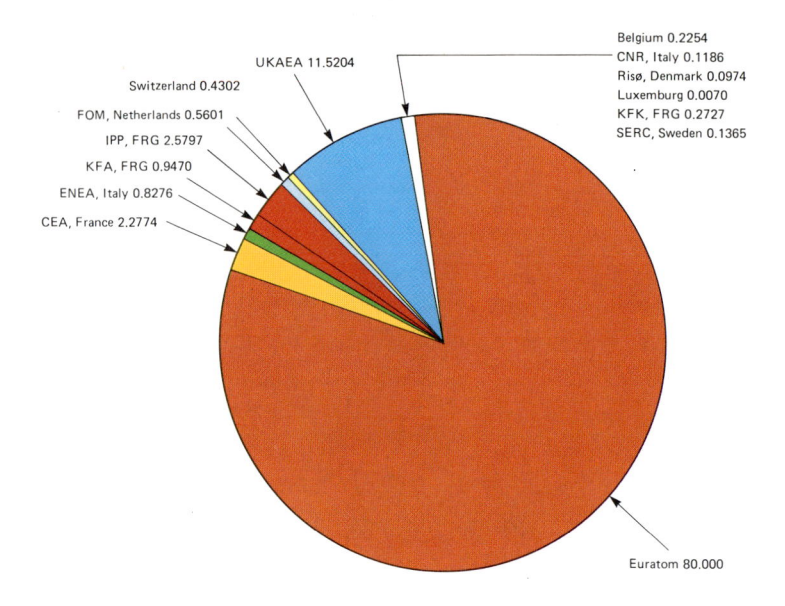

Belgium 0.2254
CNR, Italy 0.1186
Risø, Denmark 0.0974
Luxemburg 0.0070
KFK, FRG 0.2727
SERC, Sweden 0.1365

UKAEA 11.5204
Switzerland 0.4302
FOM, Netherlands 0.5601
IPP, FRG 2.5797
KFA, FRG 0.9470
ENEA, Italy 0.8276
CEA, France 2.2774

Euratom 80.000

The percentage contributions of the different Partners to the budgets of JET in 1984. The exact figures change a little from year to year, but essentially the European Communities pay 80%, to which all members plus Sweden and Switzerland contribute, the UK as host state pays 10% and, in addition, 10% is shared between all the laboratories benefiting from the joint European Fusion Programme.

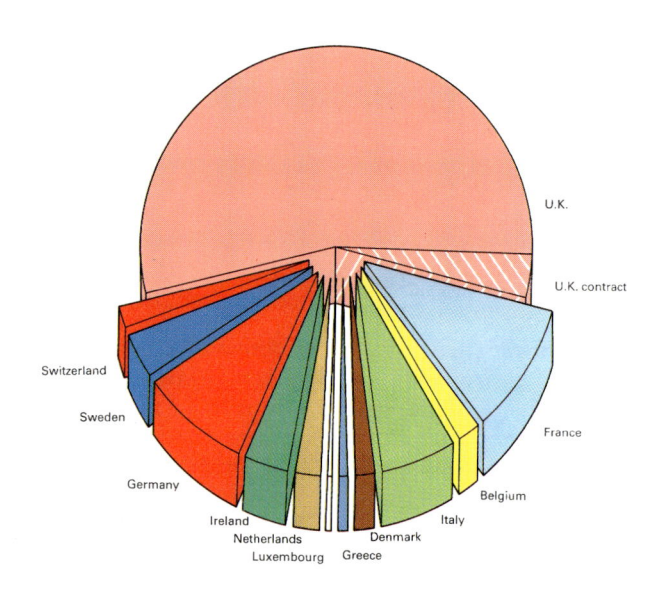

U.K.
U.K. contract
Switzerland
Sweden
Germany
Ireland
Netherlands
Luxembourg
Denmark
Greece
Italy
Belgium
France

Staff distribution by country at the transition to operation. In the higher grades, the numbers reflect more closely the strength of the fusion effort in each country.

On 18 March 1979, Dr. G. Brunner, the EC Commissioner responsible for energy research, education and science, laid the foundation stone of the JET Laboratory at the entry from the JET administration building (the responsibility of the UKAEA).

Sir John Hill (right), Chairman of the United Kingdom Atomic Energy Authority, hands over the main administration building K1 to Professor J. Teillac, President of the JET Council while Dr. R.S. Pease looks on. This was on 16 October 1980, three years less one day after the meeting of the Council of Foreign Ministers at which the structure, financing and staffing were approved.

Assembling the inner poloidal field coil from its eight component coils. This forms the primary winding of the JET transformer.

Sections of the vacuum vessel being manufactured under clean room conditions. Each octant is made up of five sections welded together through the intermediary of double walled bellows.

Assembly work in the Torus Hall began in January 1982 with the setting up of data points, the installation of the ring water main in the basement and the fixing of the eight lower horizontal beams of the transformer iron core. Positioning of the vertical limbs followed.

With all uprights in place, the lower ring and collar were mounted over the No. 2 poloidal field coil, the four (yellow) columns erected to support the upper ring and collar, the inner cylinder secured after which Nos. 3 and 4 poloidal field coils were lowered down preparatory to the insertion of the first octant.

A toroidal field coil, one of the 32 identical coils, is lowered into a one sixteenth segment of the mechanical shell structure.

Turning the segment with its two toroidal field coils in place proved to be a complex task. It took three days to turn the first, but once techniques were established, the process became routine.

Left:
Two segments mounted on either side of a vacuum vessel octant are swung round to enclose it and are bolted together to form a complete octant.

Below:
The moment of truth when the last octant was steered into the one remaining gap in the machine torus.

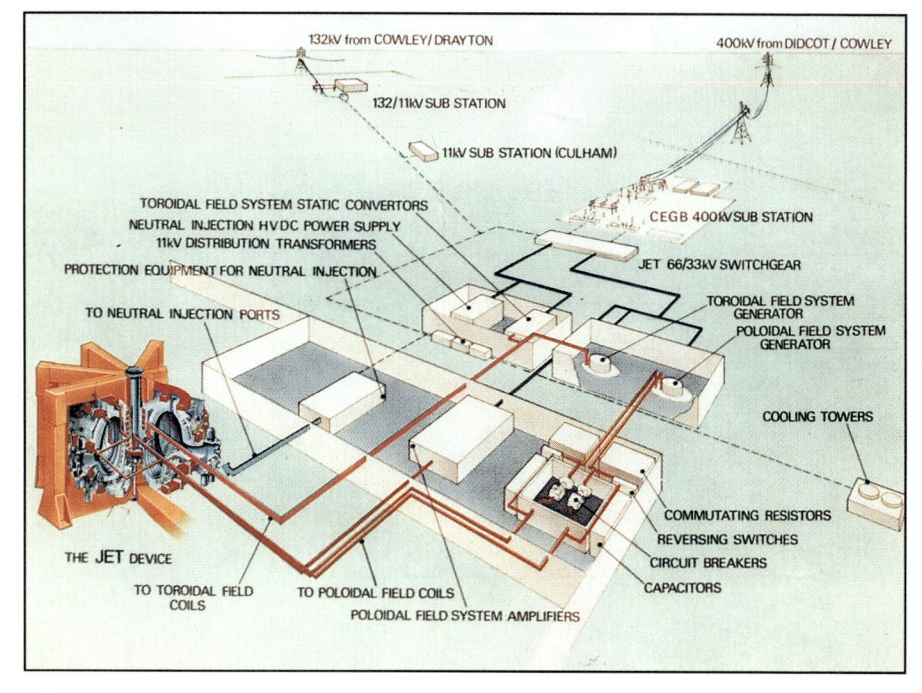

The elementary lay-out of the power supply system. The direct supply from the national grid network is supplemented at the times of peak demand by the motor generator sets where energy is stored in heavy fly-wheels.

Assembling the rotor of one of the two motor generator sets, a system weighing over 700 tons.

The JET site as seen from the South with K1 in the foreground. Beyond, the Torus Hall and main Assembly Hall to the left and right respectively of the two raised sections which allow the shielding to be lifted so that the travelling crane can pass from one to the other.

The JET site as seen from the North with, on the right, the Culham Laboratory. In the foreground the park for the static power supplies and the electricity sub-stations and on the horizon a feature of the surrounding countryside, Wittenham Clumps.

Control of the machine is centred in one room from where messages are transmitted by computer. Machine readings return by the same means. The diagnostics centre is along-side and the main computers in the room behind.

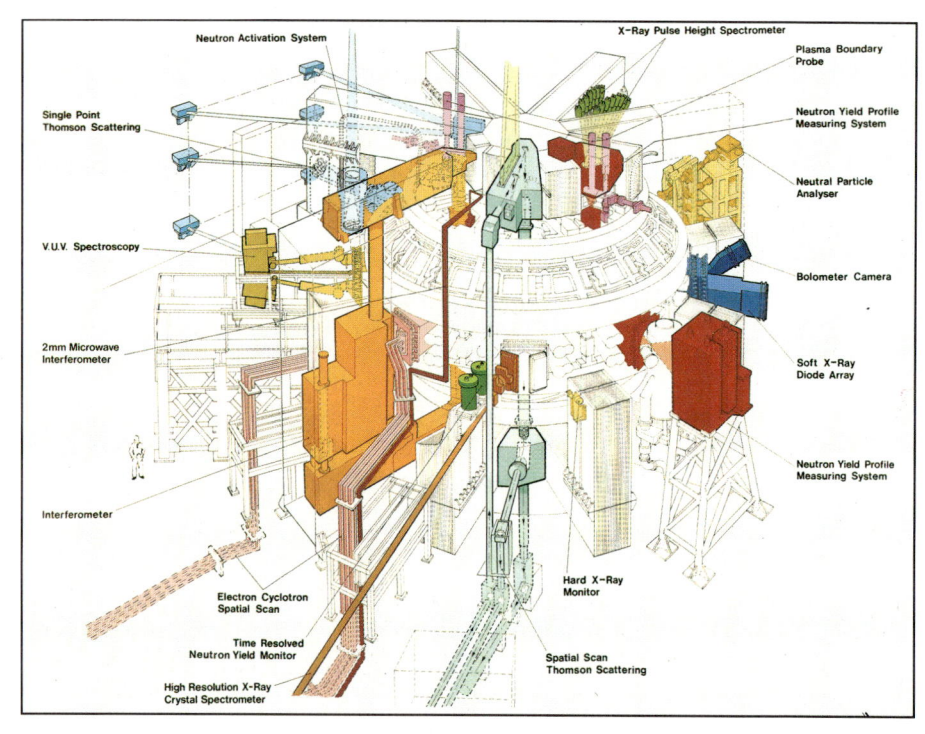

A sketch of the diagnostic systems planned to surround the torus. Over 30 instruments were being developed by the Partners although less than one third were available for initial operation.

Left:
A vacuum rotary valve of large dimensions needed to shut off the neutral injection system from the torus when not in operation. The novelty of components such as this demanded a considerable development effort on the part of the industry.

Below:
General arrangement of one of the two neutral injection auxilliary heating systems. Eight PINI sources arranged in pairs inject high energy neutral particles into the plasma. Ionisation and acceleration are followed by neutralisation and the dumping of the charged particles that remain.

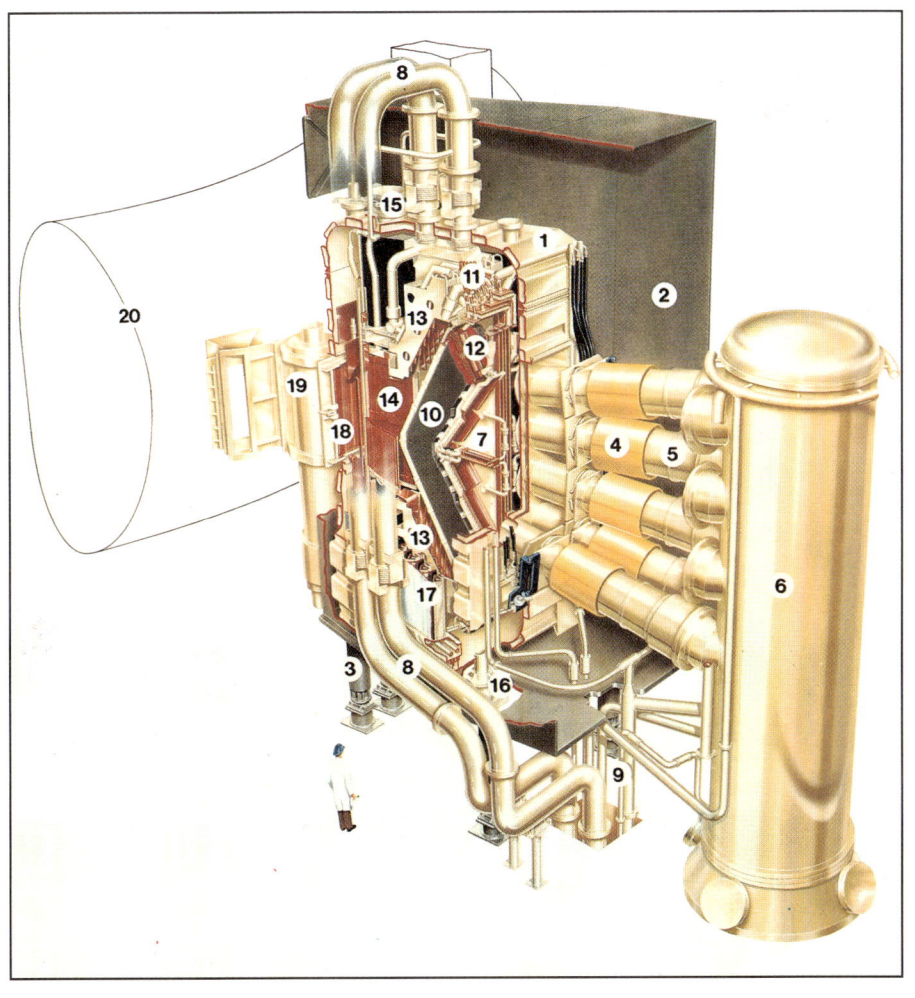

A view of the interior of the torus showing one of the two antennae used to launch RF waves into the plasma to provide auxilliary heating.

Remote handling tools have been developed to enable modifications to be made to the machine even when radioactive. Tools must be inserted through one of the entry ports after which arms unfold to reach round inside; TV cameras survey progress.

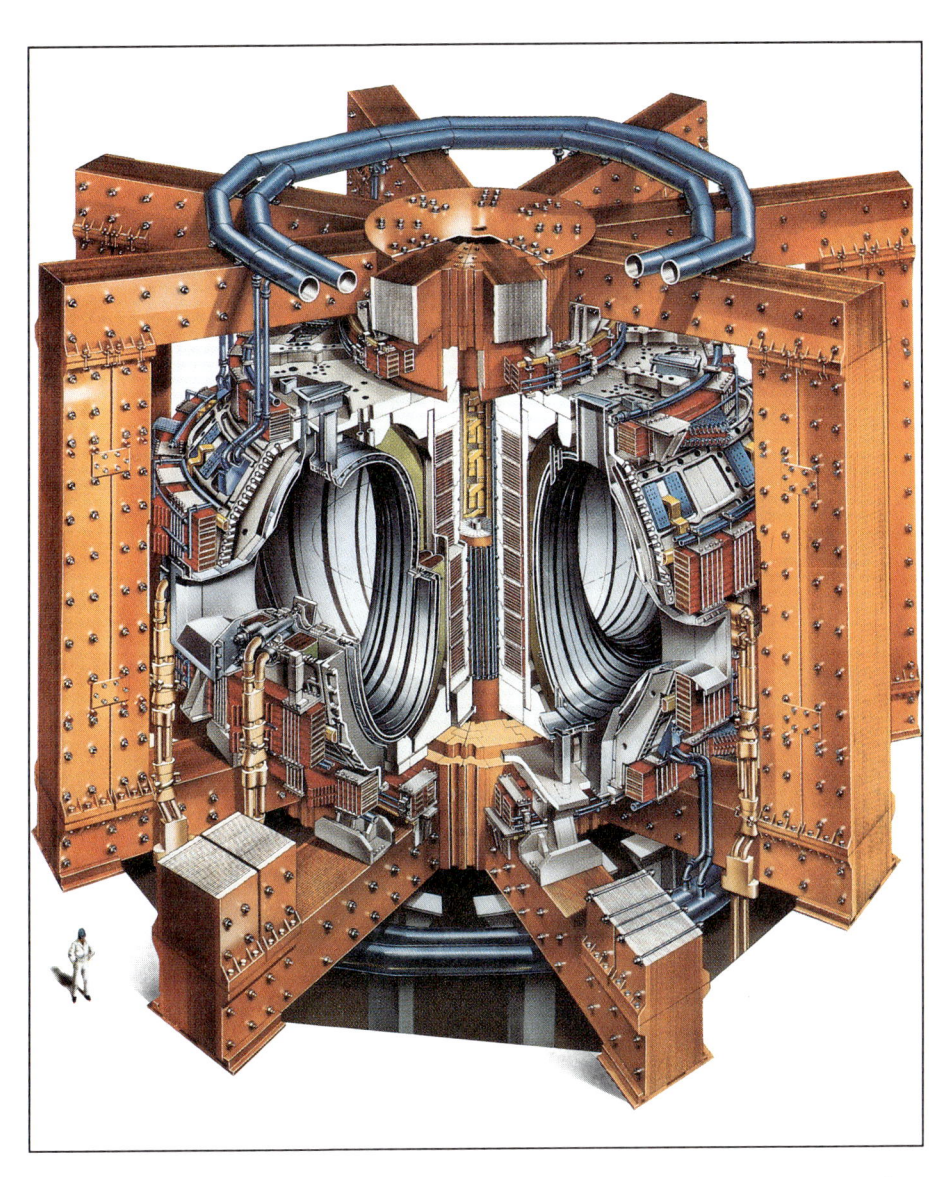

A cut-away drawing of the JET device showing the basic machine components, but none of the diagnostics or additional heating.

A photograph of the JET device – much more complicated in reality than in the preceding drawing.

Day 1 and eyes are fixed on the little Avometer balanced on top of a cupboard. In the foreground, looking unusually despondent, is Dr. H.-O. Wüster and behind him members of staff all equally anxious.

Delight as the first evidence of a plasma is examined. In the foreground, from left to right, Dr. P.-H. Rebut (back view), Dr. G. Duesing, Dr. P. Lallia, Dr. H.-O. Wüster and Dr. D. Düchs.

A view of the Assembly Hall as the staff and guests await the arrival of the royal party for the official inauguration on 9 April 1984. In the far corner can be seen the spare octant in the mounting jig.

Her Majesty Queen Elizabeth II unveils the plaque commemorating the inauguration of JET. On the platform, from left to right, Vicomte Davignon, Sir Ashley Ponsonby, Baroness Elles, H.R.H. The Duke of Edinburgh, Prof. J. Teillac, Her Majesty Queen Elizabeth II, Dr. Hans-Otto Wüster, President François Mitterrand, M.C. Thiery, M. Gaston Thorn, Mme Liliane Thorn-Petit.

Celebrations as (from left to right) Dr. P.-H. Rebut, Prof. D. Palumbo and Prof. A. Schlüter recall the long road that led to one of Europe's resounding scientific successes.

TRANSITION TO OPERATION

Long before the JET device first produced a plasma, preparations had been made for the transition from construction to operation. Fundamental to the whole question was the role that the Associations would be playing and in June 1980, Bickerton presented to the Scientific Council and the Executive Committee a paper summarizing the Project's proposals. It had been drawn up at a time when optimism was running high and the scientific programme was conceived with the 'intention to do only such physics measurements to enable the Project to proceed most rapidly to the tritium burning phase', although no use of tritium before 1986 was envisaged! In view of this and the fact that the Associations were overloaded, JET was proposing to reduce the work on diagnostics put out to them. However, JET did want to encourage participation in the experimental programme and estimated that some 80 physicists would be needed all told. Of these 40 only could be supplied from the JET staff and it was proposed the other 40 should come from the Associations. The people would be transferred to JET for periods of up to two years during which time they would be paid JET rates, half coming from the JET budgets and half from the parent laboratories.

First reactions were far from enthusiastic, Horowitz, for example, later in Council being moved to comment that the CEA and for that matter the AEA had already done more than their duty to JET as it was. Moreover at this same meeting Wüster was returning to the attack on the effects of the arbitrary cut (from 207.5 MEUA to 200 MEUA) that had been made in the construction budget in February and the need to reinstate the contingency allowance. Altogether, the Associations had the feeling that JET was becoming just too gourmand. However they agreed to consider seriously how they themselves viewed the future JET — Associations interaction and also to set down what each individual Association would like to do when the experimental programme got under way.

The situation was both novel and complex, much less amenable to the neat divisions of responsibilities possible at CERN or ILL, for example, where 'machine development' can be almost entirely divorced from the experimental programme and this, in turn, is one stage removed from operational factors. Even the basic objectives of JET could be considered ambiguous, although in the mind of the

Project Board at this time the one aim was to reach towards ignition conditions as quickly as possible, doing only such science as was necessary on the way. Others were far less sure of the wisdom of this rush and there were still, particularly at IPP, doubts about whether, for example, tritium should ever be used in JET. Moreover, the Associations had their own laboratories to think about, their own experimental programmes and many of their staff were strongly opposed to the idea of changing their way of life and getting caught up in this scientific juggernaut where science would be subordinated to a performance chase.

The Project saw it all differently. JET was the pinnacle of the Associations' endeavours, the carrier of Europe's fusion flag, and it was in the Associations' own interests to give JET first priority and, as a community, see that the investment being made was exploited to the maximum extent. The basic performance was just a first step, to be followed by enhancement programmes that would in such areas as neutral injection and power supplies, for example, be doubling the investment already being made. The coordination of this work of progressively optimizing the machine performance was clearly something that had to be done by JET under its own management and with the certainty of continuity. The problem had some resemblance to that which had faced the partners in Dragon. There each major experiment meant a new reactor configuration and the resident team considered itself to be the best qualified to plan the development of the system. This attitude led to controversy between the teams and the partners and was a major element in deciding the CEGB to postpone construction of a prototype HTGC power reactor — as it happened indefinitely. A programme committee, including representatives of the partners, similar to that which JET was proposing, had not provided a solution and for a long time had served more in Dragon as a source of dissension than an instrument of compromise.

The formal letter from JET to the Associations seeking their proposals regarding the operational phase was transmitted in November 1980 and over the next months the replies came in. All were prepared to recognise the pivotal position of JET in European fusion and wished to take an active, not subordinate, role in the experimental programme. The particular areas of interest depended on the individual laboratory. IPP's reply, signed by Wienecke's successor as Chairman of the Directorate, Klaus Pinkau, saw the role of the Associations as being naturally divided into two parts:

A   • The development and construction of hardware such as diagnostics for mounting on JET;
   • The execution at the home laboratory of experiments of particular interest to JET.

B   • The operation of equipment on JET provided by the home laboratory;
   • The execution of experiments on JET in which the laboratory played a significant role;
   • Participation in the evaluation of JET experimental results.

The reply went on to itemize the specific activities in the two categories that IPP would like to be involved in.

This sub-division was to prove the key to resolving one of the major points of difference between the Project and the Associations. Whereas JET wished the emphasis to be placed on the transfer of staff into the JET complement for long periods — preferably two years — with any technical help being provided from within JET, the Associations saw it as vital that at least equal importance be given to visits of a few months, or weeks even, where scientists would be accompanied by their own technicians. Also, having designed and built a diagnostics system, for example, an Association group was not prepared to hand it over, but wished to take charge of its commissioning and then to use it as an experimental probe, at the same time keeping in close contact with the home base. All too easily people can be lost to centralized institutions which instead of being a source of new experience, become rather a sink for existing knowledge. Information transfer was a central issue for the Associations and the original proposals regarding even access to results and rights to publication gave the impression that JET was becoming possessive.

Wüster tended to interpret the reactions as evidence of traditional parochialism, now out of date, and was very keen to get the matter settled so that he could begin restructuring the JET staff in preparation for the day. He was, however, persuaded by Pinkau not to hurry things. Discussions within the Council and its Committees as well as outside must be allowed to run their course; bulldozing through a solution was in the interest of neither the Project nor European fusion. Moreover discussions on the future structure should take place in parallel and the Associations had still to be convinced that the matrix system of divisional rows and experimental team columns that Bickerton was advocating was necessarily the best. Also the three shift working that was being contemplated providing, after the commissioning, 12 shifts weekly for experiments and development and three for maintenance, seemed to be indeed an excessively intensive programme of exploitation.

By June, JET had come to appreciate the Associations' wish to keep partly autonomous teams and the Associations in turn had come to understand the need for JET to be able to depend upon both equipment and personnel, with the result that a large measure of agreement had been reached on the broad principles and the drafting of a mobility contract for the assignment of staff by the Associations to JET could begin. The Executive Committee and then Council also approved the Director's proposals for setting up a Programme Committee adding the recommendation that it should meet four times a year rather than twice as originally put forward. It was fitting that Schuster at his last meeting as the representative of the Commission on the JET Council, should have the satisfaction of knowing that the European fusion family had found a consensus and the long-term integration of JET into the European programme was thus assured.

The contract that was finally approved by the JET Council at its first meeting in 1982 foresaw assignments for periods of between one and 24 months (with shorter periods covered by other mechanisms) of two distinct types. Where one or more Associations were to take responsibility in a defined area of work or in the operation of some particular diagnostics, a 'Task Agreement' would be drawn up with JET which defined the tasks to be performed, the estimated number and the qualifications of the people to be assigned, the duration of the assignment and their reporting responsibility. As individuals, the members of the task force would however be treated in the same way as the other category of staff who were assigned directly, and their pay and allowances would correspond. The Associations, largely on the insistence of the AEA which feared the consequences if JET should appear to be paying even part of the salary of its staff, undertook to continue to be the employers of the assigned staff and to pay their salaries and any additional expenses attributable to working on JET, the latter being reimbursed by the Commission. The Commission was in addition to pay travelling expenses and an expatriation allowance except to those of British nationality residing in the UK. Wüster's keenness to encourage long stays to preserve continuity was recognised by married staff receiving a higher expatriation allowance in the second year than in the first while removal expenses were only paid for assignments lasting longer than one year. Quite separate from these arrangements, was the provision made for a small number of Visiting Scientists from countries not represented by Associations. These are invited for periods ranging from six months to two years and enter JET via the AEA which employs them as temporary research associates in grades that relate to their standing and experience. Their salary is reimbursed to the AEA by JET on the same basis as normal British staff.

Staffing

In October 1981, starting with O'Hara, the first nominations for positions in the Operational Phase were made, followed in March 1982 by the remaining Heads of Division. In the Scientific Department the appointments paved the way to establishing in June the new structure with Gibson as Deputy Head of Department to Bickerton and in charge of a physics operations and a diagnostic engineering group, and as already noted with Engelhardt and Stott as Heads of two Experimental Divisions and Düchs as Head of Theory Division. In Rebut's Department a change of title from Construction Department to Operation and Development Department was due to take place at the end of the year. The Site and Buildings Division was to disappear and Vallone would move on. Smart was retiring (at least from full-time activity) and Poffé would become Deputy Departmental Head with the closure of the Assembly Division. For Bertolini (Power Supplies), Bombi (CODAS) and Jacquinot (RF) nothing very much would alter; Huguet was to head the Torus Division and Duesing would be concentrating on Neutral Injection. The new name at this level was that of John Dean called to head the Fusion Technology Division.

The Administrative Department would continue under O'Hara essentially as before.

Once appointed the Division Heads had then to define the staff they would be needing in the years to come. All personnel were given to understand that their present contract would be terminating in the middle of 1983 and if they wished to continue in JET they would have to apply for the posts that were being announced. Preference would be given to internal candidates, but there was no automatic transfer. Many were deeply resentful of this procedure, but as it applied to everyone, they could not complain of victimisation. Even the Directors were subject to the same rule, being formally appointed in June and Wüster took pains to explain to the staff the necessity for acting in this way and offered to meet personally anyone who felt they had a grievance or simply were in need of guidance.

Before going too far, however, the Council had to approve the staff complement and this was to prove no easy tussle. Back in October 1977 the Council of Ministers had agreed for the Construction Phase to allocate a maximum staff of 320 people of whom a maximum of 150 would fill Euratom posts. All were to be seconded from the Associations with the exception of a few from countries which belonged to the Communities but had no central fusion effort. Those figures were never achieved, and the Associations resisted any wider recruitment. The maximum attained was 272 yet when in March 1982 Wüster presented his requirement for the operational phase, there were 722 people on site. Of these 367 were JET people in terms of duties and the rest were contractor's men or people supplied by the Authority under contract. The grey group making up the difference between this number and the 255 posts (129 Euratom) actually filled at that moment worked as contract labour, and were engaged as engineers, programmers, planners and in other skilled occupations: they were not just draughtsmen over and above the number the Authority was able to provide.

With the completion of construction, many of the contractors' men would, of course, leave the site and a number of the big teams such as those employed on the buildings and on the assembly contract would be dispersed. In their place, however, would be the operations teams — working three shifts — and the teams responsible for the enhancement programme. Wüster, as a consequence, saw little diminution in the total number of personnel needed and proposed a complement that would rise to a peak of 704 people. Anticipating the reaction this would produce he also proposed to the Executive Committee that it set up a specialist Working Party to examine the data presented. This it duly agreed to do and a 6-man group was formed with Andreani from Frascati as the Chairman.

To a large extent the 'staff' content was predetermined, as already in March 1981 when preparing the proposals for the JET part of the 1982-1986 pluriannual programme, Council had approved a future staffing of 480 posts (half professionals) and had compromised on a figure of 180 Euratom posts (limited to 155 in 1982). For comparison, at this time the total number of people working in the

European fusion programme was estimated to be 3500-4000, of whom about 1000 were professionals. The difference between the 704 and 480 was to be made up in a rather unspecified way from people working on contracts with industry or with the host organisation. Although not defined as such, Andreani and his Group when they came to discuss the numbers with the different Division Heads concluded that of the total, some 150 people (40 professional) would be concerned with designing and installing the equipment necessary to bring the system to full performance, while the rest would be concerned with operations as such.

The Group's report was drawn up against a background of financial difficulties in the Communities with the Parliament blocking the budget and the Pluriannual Programme still not approved. Almost everywhere in Europe the accent was on saving rather than investment and JET's demands were considered to be excessive. On the grounds that certain skills could be pooled between Divisions, it was proposed that the total number of units be reduced by at least 50 and it was recommended there should be a minimum reduction of eight professionals, 16 B-grade and 30 C-grade. Moreover the Project should plan its programme so that the peaking in the period 1984-1986 was eased. Only Italy and Switzerland were prepared to back the Project's figures, the majority more or less followed the reports, and France even was in favour of a total of 600 only. Wüster conceded in the Executive Committee that the number of C-grade people would come down and even a few B-grade could be spared, but he defended very strongly the A-grade estimates.

In the Council when the figures were discussed in June, the opinion was that JET had to be contained and it was decided that a limit of 650 should be placed on the 1984-86 peak and a real effort made to reduce this further. Ironically, at the same time, the Project was reporting on the disappointing response from the Associations to the vigorous recruiting campaign that had been launched to fill the senior posts that had been advertised. In effect, the Associations had already been drained and more and more JET would have to look further afield while the letter of the law concerning return tickets was respected by the Associations undertaking to give a few months employment at the end of a period at JET to allow the employee time to look around for another job. By the end of 1983, of the 406 posts allocated, only 355 were filled: 100 by Euratom staff members, and 130 by AEA staff members, and the rest through contract labour. It should be noted that none of these numbers include the 200 or more people in the Culham Laboratory working *for* JET as opposed to *in* JET.

By that time, JET had a competitor at the international level. In the pluriannual programme for 1982-86 much emphasis had been placed on the Next Step, the Next European Tokamak (NET) and in March 1983 the NET-team was established by an agreement between Euratom, the Associations, Ireland, Luxembourg and Greece. JET was not a party to this initiative, it would seem by mutual agreement: the Associations did not want JET suffocating NET and JET did not want to divert effort from its own programme. NET was established on the premises of IPP and in June, Toschi resigned from his position at Frascati and from the Chairmanship of the Executive Committee to become the Director of the new project. By the end of the year NET was employing some 20 professional staff and was planned to build up to 35 over the next few months.

Within the Associations, construction had started at Garching on the ASDEX upgrade (as well as on an 'Advanced Stellerator'), at Frascati on an upgrade of the FT (currently holding the world record for the magic product $n_i . \tau_E . T_i$) and at Cadarache on a superconducting tokamak — Tore Supra. The national competition for effort was thus also considerable. To the surprise of many, Wüster had presented no opposition to these (and other) national projects going forward in competition with JET when they were discussed in the Consultative Committee of the Fusion Programme. His argument was that the international effort required a strong national base to support it and new projects were needed at the base line to keep up the momentum and provide the training for new staff as well as the additional scientific information. Assuming that JET would rapidly achieve its goals of defining the general criteria for a reacting plasma, the fusion community would have to grow considerably in strength in order to exploit the results.

Status of the British Staff

Another staffing problem was, however, rumbling in the background. The British staff, some of whom had begun working for JET in 1973, were reacting to the big pay differential between themselves and the Euratom employees and through the Staff Representative Committee raised with the JET management the whole question of the legality of the original Statutes. Whereas the Treaties setting up the Communities explicitly outlawed any discrimination in Community activities on the basis of nationality, the criterion determining whether an employee should be appointed to an AEA post rather than a Euratom post was the candidate's nationality. On the face of it, it seemed that the Statutes the Council of Ministers had approved were contravening general principles laid down in the Treaty of Rome (to which later Community members had all subscribed).

This was a situation that could not be ignored and so that both the Council and the staff could receive expert opinion on the subject, Wüster agreed that the Project would pay for legal advice. This was not a statement of sympathy with the cause, but rather an anxiety to have the matter cleared up as quickly as possible. He was fully aware of the cost implications to the Project of any change, and the problems that would arise between the UK JET staff and the rest of Culham.

In June 1983, counsel advised that indeed a *prima facie* case existed. In particular, it was considered that UK staff not previously employed by the host should have been treated as other Community nationals and that in any case all UK nationals were discriminated against because they were not allowed to compete for Community posts. Consequent upon this, the majority of the British staff agreed to pay 1% of their salary into a fighting fund to take their case to the European Court. Wüster did approach the Authority with some suggestions on token compensations from JET funds, but the Authority was adamant in its opposition.

Despite the dissatisfaction and the determination to try and alter their status, the spirit inside JET never faltered and one found the same enthusiasm, the same dedication and the same long days (and nights) being worked by the British staff as by the Euratom staff.

Footnote: After a long drawn out process with much delay being caused, for example by the Commission's attempts to reject the appeal on matters of form only, and despite favourable recommendations from successive Advocates General who upheld the claims of the appellants, the full Court found against them and on 15 January 1987 rejected the claims with costs. The staff were hugely disappointed and mildly rebellious at first. Lomer, on the other hand, it was reported, slept peacefully for the first time in years.

OPERATION AND INAUGURATION

The internal organisation of JET was designed to take account of the dual task that lay before the Project. The Scientific Department was responsible for the definition and execution of the experimental programme, the diagnostic equipment and the interpretation of results. The Operations and Development Department was responsible for the operation and maintenance of the JET device and auxiliary systems, and also for its development to full performance. Budgets reflected the two aspects, the operations budget being approved by the Council on a yearly basis, while the enhancement programme was defined in terms of the work to be done against estimates made at current prices. Both were covered by the Project Development Plan which the Council approved annually in October.

Whilst the Scientific Council advised the Council on the overall scientific policy, the newly set up Programme Committee had the task of advising the Director of JET on the specific research programme to be followed. The members were appointed *ad personam* from relevant laboratories including those in the USA and the Director chaired the Committee himself. It met four times a year and having examined proposals from both within JET and the Associations made its recommendations for the next 12 months. Planning the actual operation of the machine was the Experiments Committee, also chaired by the Director. It included Department and Division Heads and various team leaders such as those in charge of Task Agreement activities. It met every one or two weeks to decide on the following three weeks programme on the basis of proposals submitted by Programme Leaders. Rebut and Bickerton then ran their own regular meetings with the appropriate Heads of Division and Group Leaders to plan the operation and report on the results obtained.

In charge of the programme would be two Programme Leaders, one from each department, and they would typically be appointed for a run period. Each week, however, there would be a Session Leader appointed to see the programme was carried out and, and at all times during a run, there would be an engineer responsible to the Director for safety and a physicist overseeing the measurements. An interesting innovation which was readily accepted and has worked well was to appoint the Session Leaders on the basis of principal objectives regardless of rank. As a consequence the Session Leader could change from Director of Department to junior group leader from one week to the next yet both would be exercising the same authority in regard to the programme execution. The functional hierarchy was respected by all—with perhaps the exception of Rebut who, to quote one senior physicist, 'had a floating role which made it a bit difficult, as he had a tendency to meddle—but then he was so often right'.

With so many sub-systems still to be commissioned and the limited number of staff available there was no possibility of rushing into three shift operation and from September to December 1983, the main operating period of the year, the work was organized on a 14-day schedule of which four days were allocated to operation and the remainder to maintenance and commissioning. Even so, it was enough to show the Team just how much data was amassed each discharge and how difficult it was to analyse what was happening and draw logical conclusions about what to do next. Meetings of the Experiments Committee could be stormy occasions with Wüster becoming explosively exasperated by what he saw as his people's largely intuitive (i.e. non-scientific) approach to programme planning. Whereas Wüster started from his experience of beam tuning on accelerators, where parameters can be changed step-by-step according to a systematic pattern, and where the results being evaluated are largely reproducible under any set conditions, the long-serving fusion people were conditioned to devices that changed characteristics from discharge to discharge for largely unexplained reasons, and tuning owed as much to empirical instinct as to scientific reasoning. Whilst Wüster had to become acclimatized to the vagaries of tokamaks and the difficulties of carrying out exhaustive data scans with a device which could deliver only some 20 or 30 discharges per day, many of the Team had to learn the discipline of working with a device the size of JET and adopt a more systematic methodology.

Run-up

The first plasmas produced in JET had high electrical resistance—a loop voltage of say 40 volts for a current discharge of 100 kA—indicating a high level of impurities. Early on there was even chlorine left over from detergent and perspiration at the level of a fraction of a per cent, but with baking and running the glow discharge system a steady improvement was achieved. No height control of the plasma position was available when JET was switched on and even the horizontal control had still to be tuned so that only modest discharges could be tried whose duration was relatively short (¼ second) before the plasma drifted down to the bottom of the chamber or bumped into a side wall. Nevertheless in the brief period June/July of 1983, the current had been pushed up to 625 kA and the loop voltage had come down to 14 volts.

No attempt was made to use the water-cooled nickel clad limiters and these were left retracted outside the space defined by some carbon tiles that had also been installed as beam scrapers. It had been standard practice up to that time to define the plasma volume by metal plates which according to some theories would keep the outside of the plasma cool and so limit radiation losses. They were not brought into play initially because of the risk of damage from run-away electrons and, as the device could operate without them, it was felt better to continue that way rather than risk contaminating the system with heavy metals. Moreover, as time went on they proved to be a source of mechanical trouble as, one after the other, under thermal cycling they developed leaks between the water cooling and the vacuum. In the event, the decision was taken to put the cooling system under vacuum and leave the limiters retracted. Still later they were displaced by the new RF antennae.

When operation restarted in October, the bake-out temperature was raised to 270°C and the glow discharge cleaning was kept running for three days before attempting a plasma. The improvement was immediate and by the end of the month the current had been raised to 1.4 MA with a loop voltage of only 4 V. Commissioning was then completed of a more refined position control system and by the end of November a peak current of 1.9 MA had been achieved and plasma durations of about five seconds. The world records were tumbling and would be JET's for a long time to come. By Christmas the peak current was up to 3 MA, and pulse lengths of 10 seconds had been recorded with a 'flat top' of over a million amps lasting for several seconds. The Enriques pre-conditions and the basic R5 promises had been fulfilled.

One was still far from ignition of course. The loop voltage had dropped to 1 volt about, indicating a rather clean plasma, but also heralding the limit to the temperature that could be obtained from the current alone. Newly commissioned diagnostic equipment put the central electron temperature in the range 10 to 20 million degrees celsius, and the plasma density as about $2.5 \times 10^{19}/m^3$. JET was thus operating at a few per cent only of the Lawson criterion, but only six months after switch-on and with lots of power still to be coupled in and none of the additional heating yet installed, these were very early days. JET had made its debut on to the world tokamak stage in style. Star performance in the future could be guaranteed.

Inauguration

Confident of success, preparations for the official inauguration of JET had been going on for many months, and Buckingham Palace had intimated that HM the Queen would be pleased formally to open the Joint European Torus. President Mitterrand of France, as Chairman of the Council of Ministers of the European Communities, had also indicated his willingness to represent the Communities. JET was thus plunged into the intricacies of the protocols of international events.

It was hardly a happy time in the European Communities. The UK team had taken up the cudgels against the Common Agricultural Policy; France was trying to raise farm prices while fighting Britain over mutton, and Italy and Spain over wine; the UK was demanding that its payments corresponded more closely to its returns. A meeting of heads of state in Athens early in 1984 brought no break in the deadlock and the newspapers seemed to be unanimous in condemning it as a disaster. The meeting of foreign ministers which followed was inconclusive; the UK refused to continue payments pending negotiations and vetoed an increase in the VAT percentage paid to Brussels. France and Italy riposted by blocking the repayments due to the UK from previous years; the total income of the Communities was insufficient to pay the agricultural bill for more than nine months and everywhere there was talk of bankruptcy. Only the most cynical of the civil servants in Brussels classified the situation as 'business as usual'. JET was one of the few bright spots on the European horizon.

The date chosen for the inauguration was 9 April 1984 and invitations to attend were sent out to the leaders of the political and scientific communities in the name of the President of the JET Council, Jean Teillac. Problems of whether Commissioners should be 'invited' in view of JET being a venture of the European Communities as against being hosts, whether this was more a DG XII affair than one for the whole Commission, how big the party from France needed to be to ensure the President's safety . . . these and all the innumerable details were eventually resolved and the great day arrived.

The Assembly Hall had been cleared for the occasion and 1200 people were seated facing the official platform that had been erected mid-way along the wall that gave on to the active area housing the neutral injection test plant. With Her Majesty and President Mitterrand on the platform were M. Gaston Thorn, President of the Commission of the European Communities, HRH, The Duke of Edinburgh, Mme. Thorn, Baroness Elles, Vice-President of the European Parliament, Vicomte Davignon, Vice-President of the CEC and Director-General of DG XII, Sir William Heseltine, Deputy Private Secretary to Her Majesty, Sir Ashley Ponsonby, Lord Lieutenant of Oxfordshire, in addition to Teillac and Wüster.

The Opening Act which the Queen performed was to activate the mechanism opening the door from the Assembly Hall to the Active Area. Prudence dictated however that while complete confidence could be placed in the system, the door should not be closed completely and an 80cm opening had been left so that in the unthinkable event of failure, the royal party could still enter the Torus Hall. Such a precaution seemed only too wise to many of those watching as they were convinced for a long moment that after the Queen had closed the switch, nothing was happening. But finally they could see that the gap was widening and the 350 ton door slowly moved aside to their audible relief. Once again the nerves were to be strained as the Queen came to unveil the plaque marking the official opening. Having vainly looked in this high technology laboratory for an electrical device, she then went to the wrong side of the curtains to find the unveiling chords. At this point Wüster leapt to his feet and showed her the way. Unperturbed, the Queen crossed the scene, drew on the cords now revealed and the curtains solemnly parted. The media had their story and even the BBC's Overseas Service found the incident more interesting than the dozen years and more of collective human endeavour that had led up to it.

In the speeches which accompanied the opening, tribute was paid to the men and women who had conceived and built the device and to those that had created the conditions that made it possible. European countries can collaborate, Europeans can work together and when they do the results are unsurpassed anywhere in the world.

After the ceremony, albums were presented and a visit was made of the JET device and the ancillary systems. And then lunch—the official party and leaders of delegations departed for the Culham Restaurant while over 2000 sat down to eat in the marquees erected in front of building K1. JET staff and wives were interspersed amongst the guests and with every reason, JET was once again *en fête*.

After lunch the Queen left for other pressing duties while President Mitterrand walked back to JET to meet the French staff in the Council Chamber and ministers met the staff of their own countries in separate rooms. A final touch to a royal occasion came a few days later when President Mitterrand announced to Rebut that he had been named 'Chevalier de la Légion d'Honneur au titre du Ministère des Relations Extérieures', to add to his title of Chevalier de l'Ordre National de Mérite' which had been awarded in 1978.

PRINCIPAL INDUSTRIAL SUPPLIERS TO JET

ACB Nantes, F	Pumping chambers
ACM, F	Vessels, mechanical fittings
AEG Telefunken, D	Circuit breakers
Alsthom, F	Poloidal field coils
Amey Roadstone, UK	Buildings
Ansaldo, I	Ohmic heating circuit
ASEA, S	Lifting equipment
ASS, CH	Pipework, vessel inspection assembly
Astrawall, UK	Curtain walling
Balzers, FL	Vacuum equipment, PINI units and sources
BBC Mannheim, D	Toroidal and poloidal field coils, busbar system
BBC Oerlikon, CH	Poloidal field coils
BIRA Systems, USA	Control equipment
Breda, I	Inner cylinder of mechanical structure
Bruun & Sorensen, DK	Pipes and cables, instrumentation
Buderus, D	Castings for mechanical structure
CEM, F	Transformer core, busbar system
Danfysik, DK	Neutral injection bending magnets
De Pretto, I	Mechanical structures, HV components

Edwards High Vacuum, UK	High vacuum control equipment
Emile Haefely, CH	Capac system
Excel Heat, UK	Bake-out furnace
Flexider, I	Bellows for vacuum vessel
Fochi, I	Neutral injection boxes
GEC, UK	Baking plant for vacuum vessel
GEC Machines, UK	Flywheel generator convertor sets
GEC-ESL, UK	Main assembly contract
Herfurth, D	ICRF power source system
Holec, NL	AC/DC conversion unit
Interatom, D	Guide tubes and limiters
Italtrafo, I	400 kV transformers and arresters
Jordan, UK	Water cooling system
Kabelmetal, D	Copper conductors, cryotransfer lines, NI neutralisers and grids, magnet liners
L'Air Liquide, F	NI cryopumps
Leybold Heraeus, D	Turbomolecular pumps
LK-NES, DK	11 kV distribution system
MBB, D	Water-cooled grids
Merlin Gerin, F	33 kV distribution, 400 kV transformer
Merz & Mclellan, UK	Building consultants
Micropore, UK	Thermal insulation
Morfax, UK	Vacuum vessel octants
N. G. Bailey, UK	Main electrical contractor
NEI Thompson, UK	20 ton jacks
Neyrpic, F	Rotary high vacuum valves
Norsk Data, N	Codas computers and equipment
OCEM, I	PINI auxiliary power supplies
Outukumpu, SF	Copper conductors
Riva Calzoni, I	Mechanical components and civil works
Sedgwick, UK	Insurance
Sedgwick Forbes, CH,F,S,UK	Insurance

Serete, F	Pipework and water cooling
Siemens, D	PINI power supplies and prototype
SIF Bachy, F	Piling and diaphragm walls
Spinner, D	Main RF transmission line
SPL, UK	Software services
Stork Boilers, UK	Support stand and magnetic shield
Sulzer, CH	Calorimeter and ion dumps, LHE distribution
Tarmac, UK	Main building contractor and civil works
Thomsen CSF, F	Cables
Tozzi, I	415 V distribution system
Tractionel, B	Software services
VDM, D	High nickel alloys
VIP, CH	Valves

NAME INDEX

Adams, J., 6, 7, 63
Allen, A., 83, 99, 103*, 104*
Andreani, R., 33, 173
Bauer, M., 110
Baxter, J., 103
Benn, Wedgwood-, A., 64
Bertolini, E., 33, 37, 54*, 67, 110, 123, 172
Bickerton, R., 35, 135, 150, 169, 172
Bombi, F., 110, 155, 172
Booker, D., 110
Braams, C., 8*, 14, 23, 30, 46, 83, 92, 103*, 135
Brunner, G., 54, 56, 61, 71, 154
Cacaut, D., 165
Callaghan, J., 66
Celentano, G., 68*
Cockcroft, Sir John, 2, 3*
Davignon, Vicomte, 180
Dean, J., 172
Dietz, J., 165
Dokopoulos, P., 68*
Düchs, D., 151, 172
Duesing, G., 110, 146, 172
Eckhartt, D., 33, 36, 68*, 108
Eklund, S., 131
Elles, Baroness, 180
Engelhardt, W., 151, 172
Enriques, L., 21
Fox, J., 104
Fox, John, 63
Gibson, A., 17, 21, 33, 35, 36*, 68*, 110, 151, 172
von Gierke, G., 45, 86, 108
Green, B., 68*, 110, 165
Gregory, B., 63, 72
Grieger, G., 53, 54*
Hardwick, E., 48
Heisenberg, W., 6
Hill, Sir John, 157
H.M. Queen Elizabeth II, 179, 180
Horowitz, J., 11, 25, 26, 47, 58, 74, 103*, 105, 169

Huguet, M., 36*, 37, 67, 110, 165, 172
Huart, M., 163
Hurd, D., 95
Ingram, B., 165
Jacquinot, J., 148, 172
Kaufman, G., 64
Kind, P., 142
von Klitzing, G., 83, 103*, 105
de Kock, B., 165, 166
Kurchatov, I., 2

Lafleur, C., 54*, 55
Last, J., 68*, 110, 165
Lomer, W.M., 99, 176
Longo, P., 54*, 84
Luc, H., 19, 21

Maple, J., 93
Marshall, Lord, 58, 64
Massie, Sir H., 21
Matthöfer, H., 64
McMahon, J., 142
Melchinger, K.-H., 71, 83
Meusel, E.-J., 25, 26
Mitterrand, President F., 179, 180
Mondino, P.-L., 110
Morgan, P., 166

Noll, P., 44, 68*, 110, 155

Oates, P., 61, 92, 95, 103*
O'Hara, G., 96, 110, 172
Ornstein, L., 71

Palumbo, D., 7, 8*, 14, 22, 25, 26, 31, 45, 50, 51, 53, 54*, 69, 74, 83, 103*, 104*, 106
Pease, R.S., 3*, 12, 14, 20, 43, 54*, 69, 91, 99, 104*, 157
Pedini, M., 64
Pellegrini, C., 61, 110
Pinkau, K., 170
Poffé, J.-P., 33, 36*, 54*, 68*, 110, 172

Pölchen, R., 110
Ponsonby, Sir A., 162, 180
Prevot, F., 40, 45, 50

Raimondi, T., 110
Reardon, P., 68, 113
Rebut, P.-H., 15, 16*, 34, 36*, 40, 67, 68*, 103*, 109, 142, 158, 161
  Appointments, 25, 33, 74
  Design concepts, 34, 45, 47, 51, 76, 129, 147, 148
  Honours, 158, 181

Salpietro, E., 68*, 110, 161
Salvetti, C., 58, 103*
Schlüter, A., 6, 8*, 20, 23
Schmidt, Chancellor H., 66
Schmidt-Küster, W.-J., 58
Schuster, G., 26, 53, 56, 58, 69, 85, 103*, 171
Selin, K., 68*
Sheffield, J., 61, 110
Simon, J.-Y., 109
Simonet, H., 66
Smart, D., 33, 35, 36*, 110, 172
Stott, P., 151, 172

Teillac, J., 58, 70, 74, 83, 103*, 106, 157, 162, 167, 180
Thompson, E., 146
Tindemans, L., 66
Toschi, R., 33, 46, 83, 103*, 133, 175

Vallone, C., 110, 128, 165, 172
Vandenplas, P., 9, 30
Velikhov, S., 131
Venus, G., 36*, 110

Waelbroeck, F., 25
Wienecke, R., 47, 50, 65, 103*, 108
Willson, D., 18, 25, 28, 54*, 61, 92
Wüster, H.-O., 63, 72, 73, 75*, 88, 103*, 104*, 106, 108, 157, 175, 178
  Appointment, 73, 84
  Management precepts, 72, 95, 100, 105, 107, 140, 141

*Illustration

SUBJECT INDEX

Adiabatic compression, 44
Alcator, 48, 49*, 61, 132
Asdex, 46, 49*, 175
Atomic Energy Authority—see UK
Auditing, 88, 140

Barbecue, 167
Belgium, 9, 31

CEGB (Central Electricity Generating Board, 157, 162
CERN, 6, 28, 63, 72, 88, 98, 102, 123, 125
CNEN, see also Frascati, 9, 31, 106
CNR, 9, 31, 106
Commissariat à l'Energie Atomique (CEA), 7, 9, 31
Committee of Directors, 10, 18, 25
Confinement, 42, 61
Consultative Committee on Fusion, 58, 63
Contracts, 52, 77, 129, 156
Control and data acquisition, 125, 155
    Computing, 93
Costs, 30, 46, 77, 86, 134
    Indexing, 137
CREST, 57, 72
CRPP, 80
Culham, 7, 11, 91, 146
    CLT, 18, 21, 31

Danish Atomic Energy Commission (DAEC), 9, 31
DEMO, 131
Denmark, see also DAEC, 9
Design phase, 33
    Agreement, 30, 67
Diagnostics, 149
DITE, 21, 49*, 132
Divertor, 44
Dragon, 27, 52, 72, 98, 170

ENEA, 106
Enriques study, 19, 22, 34
Euratom, 1, 7, 12
    Contracts of Association, 9, 17
    JET agreements, 31, 59
    Mobility contracts, 17
    Priority projects, 17
European Communities, 12, 21, 179
    COREPER, 41, 73, 105
    Council of Ministers, 53, 61, 63, 66, 71, 73
    Fusion programme, 41, 57, 59, 131, 138, 174
    Group for Atomic Questions, 41, 52
European Centre for Medium-Range Weather Forecasts, 96, 98

Families Days, 159
Flywheel generators, 157, 162
FOM, 6, 9, 15, 31
Fontenay-aux-Roses, see also TFR, 7, 146, 148
France, see also CEA & Fontenay, 7, 11, 15, 53, 62, 73
Frascati, see also FT, 9
FT, 17, 18, 49*, 132, 175
Funding, 31, 53, 59, 74, 81

Geneva Conferences on Peaceful Uses of Atomic Energy, 3, 4
Germany, see also IPP & KFA, 6, 11, 53, 65
Groupe de Liaison, 10, 19, 22, 74
    Recommendation, 23

Heating additional, 43, 45, 61, 76, 132, 145

IAEA, 131, 133
Ignition, 43
ILL (Institut Laue-Langevin, Grenoble), 70, 98
Inauguration, 179
Industry, 40, 76, 89, 129
Intor, 133
IPP, see also Asdex, Pulsator & Zephyr, 6, 9, 31, 53

Subject Index

Views, 43, 45, 170
Ireland, 23
Ispra, 8, 53, 56, 61
Italy, see also CNEN, CNR, ENEA, 9, 15, 84

JET
  ad hoc Working Group, 25, 35
  Agreement, 52, 106
    Financial regulations, 88
    Host, 31, 91
    Staff rules, 99
    Statutes, 84
  Council, 53, 73, 85, 103*
    Interim, 83
  Device, design of, 35, 38, 41, 45, 47, 48*, 49* 112, 113,
    Assessment, 48, 60
  Executive Committee, 84
  Management Committee, 69, 83
  Programme Committee, 171, 177
  Project Board, 35, 47
  Project Control Office, 141, 153
  Reports, 38, 45, 46, 47, 67, 76, 125
  Scientific Committee, 34
  Scientific Council, 135
  Site Committee, 26, 53, 54*
  Supervisory Board, 20, 25, 31, 34, 40, 47, 50, 69
  Working Group, 20
Joint European Torus, see JET
Joint Research Centre, 8, 62,
Joint Undertaking, 71
Jutphaas, 10

Karlsruhe, 97
KFA, Jülich, 6, 31, 62

Lawson criterion, 2

Mechanical structure, 117

NET, 133, 174
Netherlands, see also FOM, Jutphaas, 6, 15, 46, 65
Neutral injection, 76, 145
New Scientist, 161

Patents, 85
Plasma, 2
  research, 5

PLT, 17, 45, 132
Poloidal field coils, 113, 114*, 121, 122*
Power supplies, see also flywheel, 29, 55, 62, 122, 123*, 157
Princeton, see also PLT & TFTR, 15, 46
Privileges, 87
Pulsator, 17, 50, 61, 132

Schooling, 94, 99
Site studies, 53
Staff, complement, 33, 47, 109, 169, 173
  Pensions, 101
  Policies, 33, 70, 84, 97
  Return ticket, 100
  Status, 97, 107, 175
Statement of Opinion, 28, 51
Sweden, 53, 79, 134
Switzerland, see also CRPP, 53, 80

T-3, 13
T-10, 49*, 132
Task Agreements, 172
Taxes, 87
Technical control documents, 142
TFR, 15, 16*, 18, 33, 49*, 50, 61, 132
TFTR, 46, 48, 49*, 61, 131
Thermonuclear fusion, 1
Tokamak, 13, 19
  Advisory Group, 17, 19, 22, 38
Tore Supra, 175
Toroidal field coils, 116*, 155, 158

UK, see also Culham & ZETA, 6, 53
  Atomic Energy Authority, 2, 9, 31, 47, 98
  Fusion Programme, 12, 21
USA, see also Princeton, 3, 11, 14, 139
USSR, 11, 13

Vacuum vessel, 119, 157
Visiting scientists, 172
Voting rights, 53, 71, 84

Wendelstein VII, 23, 28
Workshops, 45

Zephyr, 138
ZETA, 2, 3*, 13